Learn, Practice, Succ[illegible]



# Eureka Math®

## Grade 8

## Module 5

**Published by Great Minds®.**



Printed in the U.S.A.

This book may be purchased from the publisher at eureka-math.org.
5 6 7 8 9 10 LSC 26 25 24 23 22 21

ISBN 978-1-64054-984-5

G8-M5-LPS-05.2019

## Students, families, and educators:

Thank you for being part of the *Eureka Math*® community, where we celebrate the joy, wonder, and thrill of mathematics.

In *Eureka Math* classrooms, learning is activated through rich experiences and dialogue. That new knowledge is best retained when it is reinforced with intentional practice. *The Learn, Practice, Succeed* book puts in students' hands the problem sets and fluency exercises they need to express and consolidate their classroom learning and master grade-level mathematics. Once students learn and practice, they know they can succeed.

### *What is in the* Learn, Practice, Succeed *book?*

**Fluency Practice:** Our printed fluency activities utilize the format we call a Sprint. Instead of rote recall, Sprints use patterns across a sequence of problems to engage students in reasoning and to reinforce number sense while building speed and accuracy. Sprints are inherently differentiated, with problems building from simple to complex. The tempo of the Sprint provides a low-stakes adrenaline boost that increases memory and automaticity.

**Classwork:** A carefully sequenced set of examples, exercises, and reflection questions support students' in-class experiences and dialogue. Having classwork preprinted makes efficient use of class time and provides a written record that students can refer to later.

**Exit Tickets:** Students show teachers what they know through their work on the daily Exit Ticket. This check for understanding provides teachers with valuable real-time evidence of the efficacy of that day's instruction, giving critical insight into where to focus next.

**Homework Helpers and Problem Sets:** The daily Problem Set gives students additional and varied practice and can be used as differentiated practice or homework. A set of worked examples, Homework Helpers, support students' work on the Problem Set by illustrating the modeling and reasoning the curriculum uses to build understanding of the concepts the lesson addresses.

Homework Helpers and Problem Sets from prior grades or modules can be leveraged to build foundational skills. When coupled with *Affirm*®, *Eureka Math*'s digital assessment system, these Problem Sets enable educators to give targeted practice and to assess student progress. Alignment with the mathematical models and language used across *Eureka Math* ensures that students notice the connections and relevance to their daily instruction, whether they are working on foundational skills or getting extra practice on the current topic.

### *Where can I learn more about* Eureka Math *resources?*

The Great Minds® team is committed to supporting students, families, and educators with an ever-growing library of resources, available at eureka-math.org. The website also offers inspiring stories of success in the *Eureka Math* community. Share your insights and accomplishments with fellow users by becoming a *Eureka Math* Champion.

Best wishes for a year filled with "aha" moments!

Jill Diniz

Jill Diniz
Chief Academic Officer, Mathematics
Great Minds

# Contents

## Module 5: Examples of Functions from Geometry

## Example 1

Suppose a moving object travels 256 feet in 4 seconds. Assume that the object travels at a constant speed, that is, the motion of the object can be described by a linear equation. Write a linear equation in two variables to represent the situation, and use the equation to predict how far the object has moved at the four times shown.

| Number of seconds in motion ($x$) | Distance traveled in feet ($y$) |
|---|---|
| 1 | |
| 2 | |
| 3 | |
| 4 | |

## Example 2

The object, a stone, is dropped from a height of 256 feet. It takes exactly 4 seconds for the stone to hit the ground. How far does the stone drop in the first 3 seconds? What about the last 3 seconds? Can we assume constant speed in this situation? That is, can this situation be expressed using a linear equation?

| Number of seconds ($x$) | Distance traveled in feet ($y$) |
|---|---|
| 1 | |
| 2 | |
| 3 | |
| 4 | |

## Exercises 1–6

Use the table to answer Exercises 1–5.

| Number of seconds ($x$) | Distance traveled in feet ($y$) |
|---|---|
| 0.5 | 4 |
| 1 | 16 |
| 1.5 | 36 |
| 2 | 64 |
| 2.5 | 100 |
| 3 | 144 |
| 3.5 | 196 |
| 4 | 256 |

1. Name two predictions you can make from this table.

2. Name a prediction that would require more information.

3. What is the average speed of the object between 0 and 3 seconds? How does this compare to the average speed calculated over the same interval in Example 1?

$$\text{Average Speed} = \frac{\text{distance traveled over a given time interval}}{\text{time interval}}$$

4. Take a closer look at the data for the falling stone by answering the questions below.

   a. How many feet did the stone drop between 0 and 1 second?

   b. How many feet did the stone drop between 1 and 2 seconds?

   c. How many feet did the stone drop between 2 and 3 seconds?

   d. How many feet did the stone drop between 3 and 4 seconds?

   e. Compare the distances the stone dropped from one time interval to the next. What do you notice?

5. What is the average speed of the stone in each interval 0.5 second? For example, the average speed over the interval from 3.5 seconds to 4 seconds is

$$\frac{\text{distance traveled over a given time interval}}{\text{time interval}} = \frac{256-196}{4-3.5} = \frac{60}{0.5} = 120;\ 120 \text{ feet per second}$$

Repeat this process for every half-second interval. Then answer the question that follows.

   a. Interval between 0 and 0.5 second:

   b. Interval between 0.5 and 1 second:

   c. Interval between 1 and 1.5 seconds:

   d. Interval between 1.5 and 2 seconds:

e. Interval between 2 and 2.5 seconds:

f. Interval between 2.5 and 3 seconds:

g. Interval between 3 and 3.5 seconds:

h. Compare the average speed between each time interval. What do you notice?

6. Is there any pattern to the data of the falling stone? Record your thoughts below.

| **Time of interval in seconds ($t$)** | $1$ | $2$ | $3$ | $4$ |
|---|---|---|---|---|
| **Distance stone fell in feet ($y$)** | $16$ | $64$ | $144$ | $256$ |

### Lesson Summary

A *function* is a rule that assigns to each value of one quantity a single value of a second quantity. Even though we might not have a formula for that rule, we see that functions do arise in real-life situations

Name ______________________________ Date ______________

A ball is bouncing across the school yard. It hits the ground at $(0,0)$ and bounces up and lands at $(1,0)$ and bounces again. The graph shows only one bounce.

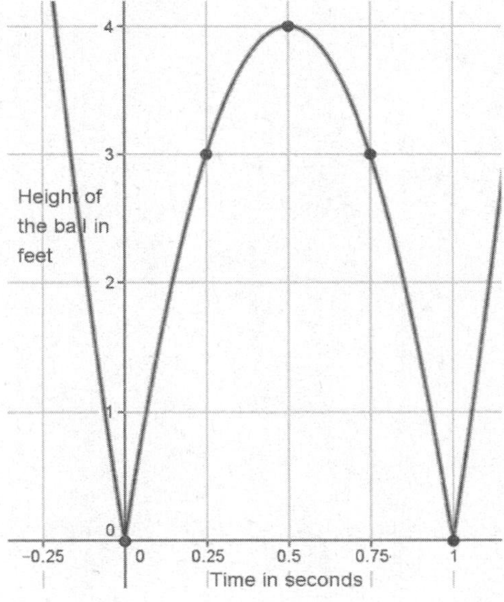


a. Identify the height of the ball at the following values of $t$: $0, 0.25, 0.5, 0.75, 1$.

b. What is the average speed of the ball over the first $0.25$ seconds? What is the average speed of the ball over the next $0.25$ seconds (from $0.25$ to $0.5$ seconds)?

c. Is the height of the ball changing at a constant rate?

Consider the path of the first 10 seconds of a roller coaster ride, graphed below on a coordinate plane. The $x$-axis represents the horizontal distance traveled, and the $y$-axis represents the height of the roller coaster.

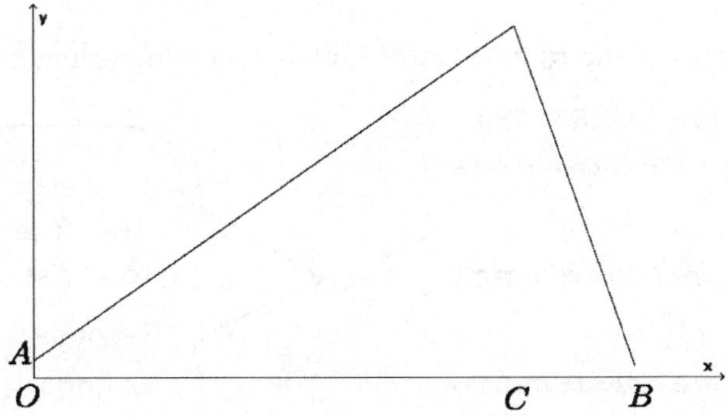


- Include the information below on the graph.
  - Point $A$ represents the platform, 20 ft. from the ground, where you get onto the ride.
  - The length of segment $OB$ is approximately 500 ft.
  - The highest point on this part of the ride, 300 ft. from the ground, is reached 8 seconds after the ride begins.

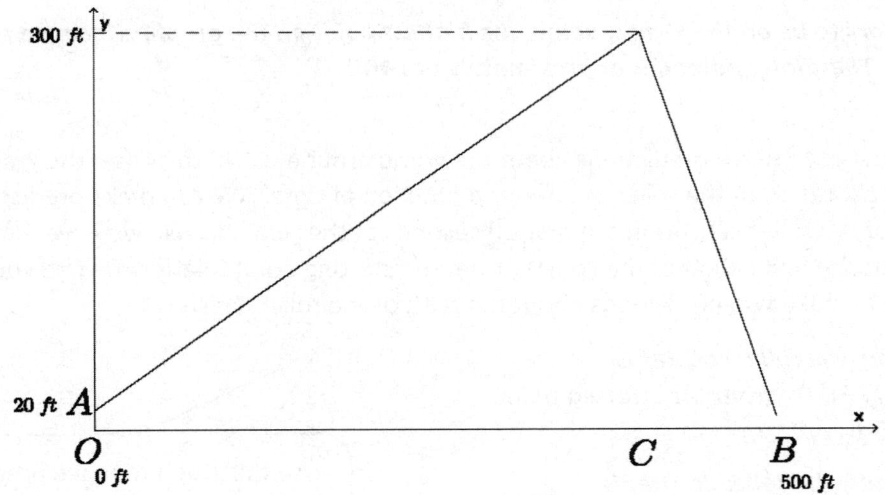


a. How much time has passed if the roller coaster is at point $A$? Explain.

***No time, $t = 0$, because this is where we get on the ride.***

b. How much time has passed as the roller coaster moves from point $A$ to point $B$?

***10 seconds. We were told that the graph represents the first 10 seconds of the ride.***

c. Approximate the coordinates of the roller coaster for the following values of $t$: $0, 2, 5, 8$, and $10$.

***At $t = 0$, the coordinates are $(0, 20)$. The coaster is on the platform, and the ride hasn't started yet.***

***At $t = 2$, the coordinates are approximately $(100, 90)$.***

I can estimate the $x$-value as 100. The reason is that the total distance traveled in 10 seconds is 500 ft., and then $\frac{2}{10}$ of 500 is 100. I estimate the $y$-value to be 90 by looking at the graph. The height after two seconds must be between 20 and 300 but closer to 20.

***At $t = 5$, the coordinates are approximately $\left(\frac{5}{10} \times 500, 200\right) = (250, 200)$.***

***At $t = 8$, the coordinates are approximately $\left(\frac{8}{10} \times 500, 300\right) = (400, 300)$. We were given the height at 8 seconds as 300 ft.***

***At $t = 10$, the coordinates are approximately $(500, 5)$.***

d. What ordered pair represents point $C$? Explain how you know.

***Point $C$ appears to be on the $x$-axis, below the highest point on the graph, which we said was at $(400, 300)$. Therefore, point $C$ is approximately at $(400, 0)$.***

e. Functions allow us to make predictions about the world around us. In this case, the graph represents the location of the roller coaster as a function of time. We can make predictions about the location of the roller coaster in the first 10 seconds of the ride because we have information about the distance and height of the coaster from the starting point (platform). Use your answers from part (d) to make two predictions about the path of the roller coaster.

***After 1 second, the roller coaster is approximately 50 ft. from the starting point and about 55 ft. up.***

***After 9 seconds, the roller coaster is approximately 450 ft. from the starting point and about 100 ft. up.***

Use the graph to make predictions about the roller coaster's distance from the starting point and its height for *any* point on the graph of its path.

A ball is thrown across the field from point $A$ to point $B$. It hits the ground at point $B$. The path of the ball is shown in the diagram below. The $x$-axis shows the horizontal distance the ball travels in feet, and the $y$-axis shows the height of the ball in feet. Use the diagram to complete parts (a)–(f).

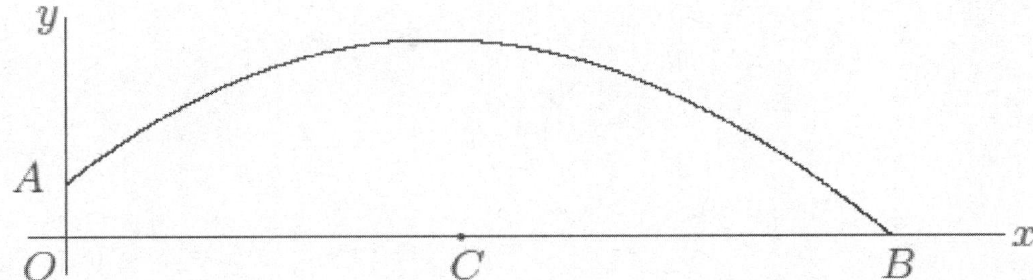


a. Suppose point $A$ is approximately 6 feet above ground and that at time $t = 0$ the ball is at point $A$. Suppose the length of $OB$ is approximately 88 feet. Include this information on the diagram.

b. Suppose that after 1 second, the ball is at its highest point of 22 feet (above point $C$) and has traveled a horizontal distance of 44 feet. What are the approximate coordinates of the ball at the following values of $t$: 0.25, 0.5, 0.75, 1, 1.25, 1.5, 1.75, and 2.

c. Use your answer from part (b) to write two predictions.

d. What is happening to the ball when it has coordinates $(88, 0)$?

e. Why do you think the ball is at point $(0, 6)$ when $t = 0$? In other words, why isn't the height of the ball 0?

f. Does the graph allow us to make predictions about the height of the ball at all points?

## Exercises 1–5

1. Let $D$ be the distance traveled in time $t$. Use the equation $D = 16t^2$ to calculate the distance the stone dropped for the given time $t$.

| Time in seconds | 0.5 | 1 | 1.5 | 2 | 2.5 | 3 | 3.5 | 4 |
|---|---|---|---|---|---|---|---|---|
| Distance stone fell in feet by that time | | | | | | | | |

a. Are the distances you calculated equal to the table from Lesson 1?

b. Does the function $D = 16t^2$ accurately represent the distance the stone fell after a given time $t$? In other words, does the function described by this rule assign to $t$ the correct distance? Explain.

2. Can the table shown below represent values of a function? Explain.

| Input ($x$) | 1 | 3 | 5 | 5 | 9 |
|---|---|---|---|---|---|
| Output ($y$) | 7 | 16 | 19 | 20 | 28 |

3. Can the table shown below represent values of a function? Explain.

| Input ($x$) | 0.5 | 7 | 7 | 12 | 15 |
|---|---|---|---|---|---|
| Output ($y$) | 1 | 15 | 10 | 23 | 30 |

4. Can the table shown below represent values of a function? Explain.

| Input ($x$) | 10 | 20 | 50 | 75 | 90 |
|---|---|---|---|---|---|
| Output ($y$) | 32 | 32 | 156 | 240 | 288 |

5. It takes Josephine 34 minutes to complete her homework assignment of 10 problems. If we assume that she works at a constant rate, we can describe the situation using a function.

   a. Predict how many problems Josephine can complete in 25 minutes.

b. Write the two-variable linear equation that represents Josephine's constant rate of work.

c. Use the equation you wrote in part (b) as the formula for the function to complete the table below. Round your answers to the hundredths place.

| **Time taken to complete problems** ($x$) | $5$ | $10$ | $15$ | $20$ | $25$ |
|---|---|---|---|---|---|
| **Number of problems completed** ($y$) | $1.47$ | | | | |

After $5$ minutes, Josephine was able to complete $1.47$ problems, which means that she was able to complete $1$ problem, then get about halfway through the next problem.

d. Compare your prediction from part (a) to the number you found in the table above.

e. Use the formula from part (b) to compute the number of problems completed when $x = -7$. Does your answer make sense? Explain.

f. For this problem, we assumed that Josephine worked at a constant rate. Do you think that is a reasonable assumption for this situation? Explain.

**Lesson Summary**

A *function* is a correspondence between a set (whose elements are called *inputs*) and another set (whose elements are called *outputs*) such that each input corresponds to one and only one output.

Sometimes the phrase *exactly one output* is used instead of *one and only one output* in the definition of function (they mean the same thing). Either way, it is this fact, that there is one and only one output for each input, which makes functions predictive when modeling real life situations.

Furthermore, the correspondence in a function is often given by a *rule* (or *formula*). For example, the output is equal to the number found by substituting an input number into the variable of a one-variable expression and evaluating.

Functions are sometimes described as an *input–output machine.* For example, given a function $D$, the input is time $t$, and the output is the distance traveled in $t$ seconds.

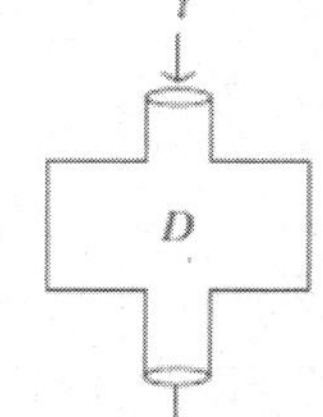

Distance traveled in $t$ seconds

Name __________________________________________ Date ____________________

1. Can the table shown below represent values of a function? Explain.

| **Input** ($x$) | 10 | 20 | 30 | 40 | 50 |
|---|---|---|---|---|---|
| **Output** ($y$) | 32 | 64 | 96 | 64 | 32 |

2. Kelly can tune 4 cars in 3 hours. If we assume he works at a constant rate, we can describe the situation using a function.

   a. Write the function that represents Kelly's constant rate of work.

   b. Use the function you wrote in part (a) as the formula for the function to complete the table below. Round your answers to the hundredths place.

| **Time spent tuning cars** ($x$) | 2 | 3 | 4 | 6 | 7 |
|---|---|---|---|---|---|
| **Number of cars tuned up** ($y$) | | | | | |

c. Kelly works 8 hours per day. According to this work, how many cars will he finish tuning at the end of a shift?

d. For this problem, we assumed that Kelly worked at a constant rate. Do you think that is a reasonable assumption for this situation? Explain.

1. The table below represents the number of minutes Esmeralda reads each day for a week. Do the data shown below represent values of a function? Explain.

| Day ($x$) | 1 | 2 | 3 | 4 | 5 | 6 | 7 |
|---|---|---|---|---|---|---|---|
| Time in Minutes ($y$) | 85 | 30 | 60 | 30 | 15 | 80 | 10 |

***Each input has exactly one output; therefore, these data represent a function.***

Check each $x$-value (input) to make sure each has only one unique $y$-value (output).

2. The table below represents the total number of steps that Jamar has taken for various months in the past two years. Examine the data in the table below, and determine whether or not they could represent a function. Explain.

| Month ($x$) | March | July | March | February | June | October |
|---|---|---|---|---|---|---|
| Number of Steps ($y$) | 215,760 | 235,842 | 201,388 | 197,094 | 220,972 | 200,578 |

***These data cannot represent a function because there are two values given for the month of March. Jamar cannot take $215,760$ and $201,388$ steps in the same month.***

I just substitute the $x$-value into the equation $y = x^2 - 1$. The answer I get, $y$, is the output.

3. A function can be described by the rule $y = x^2 - 1$. Determine the corresponding output for each given input.

| Input ($x$) | $-3$ | $-1$ | $0$ | $1$ | $3$ |
|---|---|---|---|---|---|
| Output ($y$) | $(-3)^2 - 1$<br>$= 9 - 1$<br>$= 8$ | $(-1)^2 - 1$<br>$= 1 - 1$<br>$= 0$ | $0^2 - 1$<br>$= 0 - 1$<br>$= -1$ | $1^2 - 1$<br>$= 1 - 1$<br>$= 0$ | $3^2 - 1$<br>$= 9 - 1$<br>$= 8$ |

4. Examine the data in the table below. The inputs represent the number of cans of corn purchased, and the outputs represent the cost. Determine the cost of one can of corn, assuming the price per can is the same no matter how many cans are purchased. Then, complete the table.

| Cans of Corn ($x$) | 1 | 2 | 3 | 4 | 5 | 6 | 7 |
|---|---|---|---|---|---|---|---|
| Cost in Dollars ($y$) | **$0.50$** | **$1.00$** | 1.50 | **$2.00$** | 2.50 | **$3.00$** | **$3.50$** |

a. Write the rule that describes the function.

***Let y be the cost of $x$ cans of corn.***

***Write each part of the proportion as*** $\frac{\textbf{cost}}{\textbf{cans of corn}}$.

$$\frac{y}{x} = \frac{1.50}{3}$$

$$y = \frac{1.50}{3}x$$

$$y = 0.50x$$

***Now, this problem is just like the last one!***

First, I define my variables $x$ and $y$. Then, I write a proportion using data from the table and the variables. Once I solve the proportion for $y$, I'll have the rule.

b. Can you determine the value of the output for an input of $x = -10$? If so, what is it?

***Yes, just substitute $-10$ in for $x$ to determine the output.***

$$y = 0.50x$$

$$y = 0.50(-10)$$

$$y = -5.00$$

c. Does an input of $-10$ make sense in this situation? Explain.

***An input of $-10$ means that $-10$ cans of corn were purchased. Only a positive number of cans can be purchased, so no, an input of $-10$ does not make sense in this situation.***

1. The table below represents the number of minutes Francisco spends at the gym each day for a week. Does the data shown below represent values of a function? Explain.

| Day ($x$) | 1 | 2 | 3 | 4 | 5 | 6 | 7 |
|---|---|---|---|---|---|---|---|
| Time in minutes ($y$) | 35 | 45 | 30 | 45 | 35 | 0 | 0 |

2. Can the table shown below represent values of a function? Explain.

| Input ($x$) | 9 | 8 | 7 | 8 | 9 |
|---|---|---|---|---|---|
| Output ($y$) | 11 | 15 | 19 | 24 | 28 |

3. Olivia examined the table of values shown below and stated that a possible rule to describe this function could be $y = -2x + 9$. Is she correct? Explain.

| Input ($x$) | $-4$ | 0 | 4 | 8 | 12 | 16 | 20 | 24 |
|---|---|---|---|---|---|---|---|---|
| Output ($y$) | 17 | 9 | 1 | $-7$ | $-15$ | $-23$ | $-31$ | $-39$ |

4. Peter said that the set of data in part (a) describes a function, but the set of data in part (b) does not. Do you agree? Explain why or why not.

a.

| Input ($x$) | 1 | 2 | 3 | 4 | 5 | 6 | 7 | 8 |
|---|---|---|---|---|---|---|---|---|
| Output ($y$) | 8 | 10 | 32 | 6 | 10 | 27 | 156 | 4 |

b.

| Input ($x$) | $-6$ | $-15$ | $-9$ | $-3$ | $-2$ | $-3$ | 8 | 9 |
|---|---|---|---|---|---|---|---|---|
| Output ($y$) | 0 | $-6$ | 8 | 14 | 1 | 2 | 11 | 41 |

5. A function can be described by the rule $y = x^2 + 4$. Determine the corresponding output for each given input.

| Input ($x$) | $-3$ | $-2$ | $-1$ | 0 | 1 | 2 | 3 | 4 |
|---|---|---|---|---|---|---|---|---|
| Output ($y$) | | | | | | | | |

6. Examine the data in the table below. The inputs and outputs represent a situation where constant rate can be assumed. Determine the rule that describes the function.

| Input $(x)$ | $-1$ | $0$ | $1$ | $2$ | $3$ | $4$ | $5$ | $6$ |
|---|---|---|---|---|---|---|---|---|
| Output $(y)$ | $3$ | $8$ | $13$ | $18$ | $23$ | $28$ | $33$ | $38$ |

7. Examine the data in the table below. The inputs represent the number of bags of candy purchased, and the outputs represent the cost. Determine the cost of one bag of candy, assuming the price per bag is the same no matter how much candy is purchased. Then, complete the table.

| Bags of candy $(x)$ | $1$ | $2$ | $3$ | $4$ | $5$ | $6$ | $7$ | $8$ |
|---|---|---|---|---|---|---|---|---|
| Cost in Dollars $(y)$ | | | | $5.00$ | $6.25$ | | | $10.00$ |

a. Write the rule that describes the function.

b. Can you determine the value of the output for an input of $x = -4$? If so, what is it?

c. Does an input of $-4$ make sense in this situation? Explain.

8. Each and every day a local grocery store sells 2 pounds of bananas for $\$1.00$. Can the cost of 2 pounds of bananas be represented as a function of the day of the week? Explain.

9. Write a brief explanation to a classmate who was absent today about why the table in part (a) is a function and the table in part (b) is not.

a.

| Input $(x)$ | $-1$ | $-2$ | $-3$ | $-4$ | $4$ | $3$ | $2$ | $1$ |
|---|---|---|---|---|---|---|---|---|
| Output $(y)$ | $81$ | $100$ | $320$ | $400$ | $400$ | $320$ | $100$ | $81$ |

b.

| Input $(x)$ | $1$ | $6$ | $-9$ | $-2$ | $1$ | $-10$ | $8$ | $14$ |
|---|---|---|---|---|---|---|---|---|
| Output $(y)$ | $2$ | $6$ | $-47$ | $-8$ | $19$ | $-2$ | $15$ | $31$ |

## Example 1

In the last lesson, we looked at several tables of values showing the inputs and outputs of functions. For instance, one table showed the costs of purchasing different numbers of bags of candy:

| **Bags of candy ($x$)** | 1 | 2 | 3 | 4 | 5 | 6 | 7 | 8 |
|---|---|---|---|---|---|---|---|---|
| **Cost in Dollars ($y$)** | 1.25 | 2.50 | 3.75 | 5.00 | 6.25 | 7.50 | 8.75 | 10.00 |

## Example 2

Walter walks at a constant speed of 8 miles every 2 hours. Describe a linear function for the number of miles he walks in $x$ hours. What is a reasonable range of $x$-values for this function?

## Example 3

Veronica runs at a constant speed. The distance she runs is a function of the time she spends running. The function has the table of values shown below.

| **Time in minutes** **($x$)** | $8$ | $16$ | $24$ | $32$ |
|---|---|---|---|---|
| **Distance run in miles** **($y$)** | $1$ | $2$ | $3$ | $4$ |

## Example 4

Water flows from a faucet into a bathtub at the constant rate of $7$ gallons of water pouring out every $2$ minutes. The bathtub is initially empty, and its plug is in. Determine the rule that describes the volume of water in the tub as a function of time. If the tub can hold $50$ gallons of water, how long will it take to fill the tub?

Now assume that you are filling the same $50$-gallon bathtub with water flowing in at the constant rate of $3.5$ gallons per minute, but there were initially $8$ gallons of water in the tub. Will it still take about $14$ minutes to fill the tub?

| **Time in minutes** **($x$)** | $0$ | $3$ | $6$ | $9$ | $12$ |
|---|---|---|---|---|---|
| **Total volume in tub in gallons** **($y$)** | | | | | |

## Example 5

Water flows from a faucet at a constant rate. Assume that 6 gallons of water are already in a tub by the time we notice the faucet is on. This information is recorded in the first column of the table below. The other columns show how many gallons of water are in the tub at different numbers of minutes since we noticed the running faucet.

| **Time in minutes** $(x)$ | 0 | 3 | 5 | 9 |
|---|---|---|---|---|
| **Total volume in tub in gallons** $(y)$ | 6 | 9.6 | 12 | 16.8 |

## Exercises 1–3

1. Hana claims she mows lawns at a constant rate. The table below shows the area of lawn she can mow over different time periods.

| **Number of minutes** $(x)$ | 5 | 20 | 30 | 50 |
|---|---|---|---|---|
| **Area mowed in square feet** $(y)$ | 36 | 144 | 216 | 360 |

a. Is the data presented consistent with the claim that the area mowed is a linear function of time?

b. Describe in words the function in terms of area mowed and time.

c. At what rate does Hana mow lawns over a 5-minute period?

d. At what rate does Hana mow lawns over a 20-minute period?

e. At what rate does Hana mow lawns over a 30-minute period?

f. At what rate does Hana mow lawns over a 50-minute period?

g. Write the equation that describes the area mowed, $y$, in square feet, as a linear function of time, $x$, in minutes.

h. Describe any limitations on the possible values of $x$ and $y$.

i. What number does the function assign to $x = 24$? That is, what area of lawn can be mowed in $24$ minutes?

j. According to this work, how many minutes would it take to mow an area of $400$ square feet?

2. A linear function has the table of values below. The information in the table shows the total volume of water, in gallons, that flows from a hose as a function of time, the number of minutes the hose has been running.

| **Time in minutes** **($x$)** | $10$ | $25$ | $50$ | $70$ |
|---|---|---|---|---|
| **Total volume of water in gallons** **($y$)** | $44$ | $110$ | $220$ | $308$ |

a. Describe the function in terms of volume and time.

b. Write the rule for the volume of water in gallons, $y$, as a linear function of time, $x$, given in minutes.

c. What number does the function assign to $250$? That is, how many gallons of water flow from the hose during a period of $250$ minutes?

d. The average swimming pool holds about $17{,}300$ gallons of water. Suppose such a pool has already been filled one quarter of its volume. Write an equation that describes the volume of water in the pool if, at time $0$ minutes, we use the hose described above to start filling the pool.

e. Approximately how many hours will it take to finish filling the pool?

3. Recall that a linear function can be described by a rule in the form of $y = mx + b$, where $m$ and $b$ are constants. A particular linear function has the table of values below.

| **Input** ($x$) | $0$ | $4$ | $10$ | $11$ | $15$ | $20$ | $23$ |
|---|---|---|---|---|---|---|---|
| **Output** ($y$) | $4$ | $24$ | $54$ | $59$ | | | |

a. What is the equation that describes the function?

b. Complete the table using the rule.

**Lesson Summary**

A linear equation $y = mx + b$ describes a rule for a function. We call any function defined by a linear equation a linear function.

Problems involving a constant rate of change or a proportional relationship can be described by linear functions.

Name ______________________________ Date ______________

The information in the table shows the number of pages a student can read in a certain book as a function of time in minutes spent reading. Assume a constant rate of reading.

| Time in minutes ($x$) | 2 | 6 | 11 | 20 |
|---|---|---|---|---|
| Total number of pages read in a certain book ($y$) | 7 | 21 | 38.5 | 70 |

a. Write the equation that describes the total number of pages read, $y$, as a linear function of the number of minutes, $x$, spent reading.

b. How many pages can be read in 45 minutes?

c. A certain book has 396 pages. The student has already read $\frac{3}{8}$ of the pages and now picks up the book again at time $x = 0$ minutes. Write the equation that describes the total number of pages of the book read as a function of the number of minutes of further reading.

d. Approximately how much time, in minutes, will it take to finish reading the book?

1. A particular linear function has the table of values below.

| Input ($x$) | $-2$ | $4$ | $6$ | $12$ | $15$ | $16$ | $19$ |
|---|---|---|---|---|---|---|---|
| Output ($y$) | $\mathbf{-14}$ | $16$ | $26$ | $\mathbf{56}$ | $71$ | $\mathbf{76}$ | $91$ |

a. What is the equation that describes the function?

***Any pair of data can be used to determine the rate of change. We will use $(4, 16)$ as $(x_1, y_1)$ and $(6, 26)$ as $(x_2, y_2)$.***

To write the equation, I must first determine the rate of change, $m$. Next, I will substitute the values of $x, y$, and $m$ into the linear equation $y = mx + b$ to determine the value of $b$.

$$m = \frac{y_2 - y_1}{x_2 - x_1} = \frac{26 - 16}{6 - 4} = \frac{10}{2} = 5$$

***To write the equation, we select a pair of data and use $m = 5$ in the equation $y = mx + b$.***

***We can select any input to use as $x$, but we must use the corresponding output as the $y$-value. Using $(4, 16)$ will work because $16$ corresponds to the input of $4$. Using $(4, 71)$ will not work because $71$ corresponds to the input of $15$, not $4$.***

$$y = mx + b$$
$$16 = 5(4) + b$$
$$16 = 20 + b$$
$$16 - 20 = 20 - 20 + b$$
$$-4 = 0 + b$$
$$-4 = b$$

***The equation that describes this function is $y = 5x + (-4)$ or the equivalent equation $y = 5x - 4$.***

b. Complete the table using the rule (the equation you wrote in part (a)).

***We need the outputs that correspond to the inputs of* $-2$, $12$, *and* $16$. *To determine them, substitute the value of each input,* $x$, *into the equation we found in part (a).***

***For*** $x = -2$:

$$y = 5x - 4$$
$$y = 5(-2) - 4$$
$$y = -10 - 4$$
$$y = -14$$

***For*** $x = 12$:

$$y = 5x - 4$$
$$y = 5(12) - 4$$
$$y = 60 - 4$$
$$y = 56$$

***For*** $x = 16$:

$$y = 5x - 4$$
$$y = 5(16) - 4$$
$$y = 80 - 4$$
$$y = 76$$

2. A linear function has the table of values below. The data in the table shows the time, in hours, that a car travels and the corresponding distance traveled in miles. Assume the car travels at a constant speed.

| **Number of Hours Traveled ($x$)** | 1.75 | 3.25 | 4 |
|---|---|---|---|
| **Distance in Miles ($y$)** | 101.5 | 188.5 | 232 |

a. Describe the function in terms of distance and time.

***The distance the car travels is a function of the time it spends traveling.***

Is distance a function of the time the car travels, or is time a function of the distance traveled? I normally say that the output is a function of the input.

b. Write the rule that represents the linear function that describes the distance traveled in miles, $y$, in $x$ hours.

***This is just like part (a) of the previous problem!***

$$m = \frac{232 - 188.5}{4 - 3.25} = \frac{43.5}{0.75} = 58$$

***Using*** $(x, y) = (4, 232)$ ***and*** $m = 58$ ***in the equation*** $y = mx + b$:

$$232 = 58(4) + b$$
$$232 = 232 + b$$
$$232 - 232 = 232 - 232 + b$$
$$0 = b$$

***The equation for this function is*** $y = 58x$.

1. A food bank distributes cans of vegetables every Saturday. The following table shows the total number of cans they have distributed since the beginning of the year. Assume that this total is a linear function of the number of weeks that have passed.

| **Number of weeks** ($x$) | 1 | 12 | 20 | 45 |
|---|---|---|---|---|
| **Total number of cans of vegetables distributed** ($y$) | 180 | 2,160 | 3,600 | 8,100 |

 a. Describe the function being considered in words.
 b. Write the linear equation that describes the total number of cans handed out, $y$, in terms of the number of weeks, $x$, that have passed.
 c. Assume that the food bank wants to distribute 20,000 cans of vegetables. How long will it take them to meet that goal?
 d. The manager had forgotten to record that they had distributed 35,000 cans on January 1. Write an adjusted linear equation to reflect this forgotten information.
 e. Using your function in part (d), determine how long in years it will take the food bank to hand out 80,000 cans of vegetables.

2. A linear function has the table of values below. It gives the number of miles a plane travels over a given number of hours while flying at a constant speed.

| **Number of hours traveled** ($x$) | 2.5 | 4 | 4.2 |
|---|---|---|---|
| **Distance in miles** ($y$) | 1,062.5 | 1,700 | 1,785 |

 a. Describe in words the function given in this problem.
 b. Write the equation that gives the distance traveled, $y$, in miles, as a linear function of the number of hours, $x$, spent flying.
 c. Assume that the airplane is making a trip from New York to Los Angeles, which is a journey of approximately 2,475 miles. How long will it take the airplane to get to Los Angeles?
 d. If the airplane flies for 8 hours, how many miles will it cover?

3. A linear function has the table of values below. It gives the number of miles a car travels over a given number of hours.

| **Number of hours traveled** ($x$) | 3.5 | 3.75 | 4 | 4.25 |
|---|---|---|---|---|
| **Distance in miles** ($y$) | 203 | 217.5 | 232 | 246.5 |

a. Describe in words the function given.

b. Write the equation that gives the distance traveled, in miles, as a linear function of the number of hours spent driving.

c. Assume that the person driving the car is going on a road trip to reach a location 500 miles from her starting point. How long will it take the person to get to the destination?

4. A particular linear function has the table of values below.

| **Input** ($x$) | 2 | 3 | 8 | 11 | 15 | 20 | 23 |
|---|---|---|---|---|---|---|---|
| **Output** ($y$) | 7 | 10 | | 34 | | 61 | |

a. What is the equation that describes the function?

b. Complete the table using the rule.

5. A particular linear function has the table of values below.

| **Input** ($x$) | 0 | 5 | 8 | 13 | 15 | 18 | 21 |
|---|---|---|---|---|---|---|---|
| **Output** ($y$) | 6 | 11 | 14 | | 21 | | |

a. What is the rule that describes the function?

b. Complete the table using the rule.

## Example 1

Classify each of the functions described below as either discrete or not discrete.

a) The function that assigns to each whole number the cost of buying that many cans of beans in a particular grocery store.

b) The function that assigns to each time of day one Wednesday the temperature of Sammy's fever at that time.

c) The function that assigns to each real number its first digit.

d) The function that assigns to each day in the year 2015 my height at noon that day.

e) The function that assigns to each moment in the year 2015 my height at that moment.

f) The function that assigns to each color the first letter of the name of that color.

g) The function that assigns the number $23$ to each and every real number between $20$ and $30.6$.

h) The function that assigns the word YES to every yes/no question.

i) The function that assigns to each height directly above the North Pole the temperature of the air at that height right at this very moment.

## Example 2

Water flows from a faucet into a bathtub at a constant rate of 7 gallons of water every 2 minutes. Regard the volume of water accumulated in the tub as a function of the number of minutes the faucet has been on. Is this function discrete or not discrete?

## Example 3

You have just been served freshly made soup that is so hot that it cannot be eaten. You measure the temperature of the soup, and it is $210°\text{F}$. Since $212°\text{F}$ is boiling, there is no way it can safely be eaten yet. One minute after receiving the soup, the temperature has dropped to $203°\text{F}$. If you assume that the rate at which the soup cools is constant, write an equation that would describe the temperature of the soup over time.

## Example 4

Consider the function that assigns to each of nine baseball players, numbered 1 through 9, his height. The data for this function is given below. Call the function $G$.

| Player Number | Height |
|---|---|
| $1$ | $5'11''$ |
| $2$ | $5'4''$ |
| $3$ | $5'9''$ |
| $4$ | $5'6''$ |
| $5$ | $6'3''$ |
| $6$ | $6'8''$ |
| $7$ | $5'9''$ |
| $8$ | $5'10''$ |
| $9$ | $6'2''$ |

## Exercises 1–3

1. At a certain school, each bus in its fleet of buses can transport 35 students. Let $B$ be the function that assigns to each count of students the number of buses needed to transport that many students on a field trip.

When Jinpyo thought about the situation, he drew the following table of values and wrote the formula $B = \frac{x}{35}$. Here $x$ is the count of students, and $B$ is the number of buses needed to transport that many students. He concluded that $B$ is a linear function.

| **Number of students ($x$)** | $35$ | $70$ | $105$ | $140$ |
|---|---|---|---|---|
| **Number of buses ($B$)** | $1$ | $2$ | $3$ | $4$ |

Alicia looked at Jinpyo's work and saw no errors with his arithmetic. But she said that the function is not actually linear.

a. Alicia is right. Explain why $B$ is not a linear function.

b. Is $B$ a discrete function?

2. A linear function has the table of values below. It gives the costs of purchasing certain numbers of movie tickets.

| **Number of tickets** $(x)$ | $3$ | $6$ | $9$ | $12$ |
|---|---|---|---|---|
| **Total cost in dollars** $(y)$ | $27.75$ | $55.50$ | $83.25$ | $111.00$ |

a. Write the linear function that represents the total cost, $y$, for $x$ tickets purchased.

b. Is the function discrete? Explain.

c. What number does the function assign to $4$? What do the question and your answer mean?

3. A function produces the following table of values.

| **Input** | **Output** |
|---|---|
| Banana | B |
| Cat | C |
| Flippant | F |
| Oops | O |
| Slushy | S |

a. Make a guess as to the rule this function follows. Each input is a word from the English language.

b. Is this function discrete?

**Lesson Summary**

Functions are classified as either discrete or not discrete.

Discrete functions admit only individually separate input values (such as whole numbers of students, or words of the English language). Functions that are not discrete admit any input value within a range of values (fractional values, for example).

Functions that describe motion or smooth changes over time, for example, are typically not discrete.

Name ______________________________ Date ______________

1. The table below shows the costs of purchasing certain numbers of tablets. We can assume that the total cost is a linear function of the number of tablets purchased.

| **Number of tablets ($x$)** | 17 | 22 | 25 |
|---|---|---|---|
| **Total cost in dollars ($y$)** | 10,183.00 | 13,178.00 | 14,975.00 |

   a. Write an equation that describes the total cost, $y$, as a linear function of the number, $x$, of tablets purchased.

   b. Is the function discrete? Explain.

   c. What number does the function assign to 7? Explain.

2. A function $C$ assigns to each word in the English language the number of letters in that word. For example, $C$ assigns the number 6 to the word *action*.

   a. Give an example of an input to which $C$ would assign the value 3.

   b. Is $C$ a discrete function? Explain.

1. A function has the table of values to the right that shows the total cost for a certain number of football tickets purchased.

| Number of Tickets ($x$) | Total Cost in Dollars ($y$) |
|---|---|
| $3$ | $18.75$ |
| $7$ | $43.75$ |
| $8$ | $50$ |
| $15$ | $93.75$ |

a. Is the function a linear function? Explain.

***Sample Student Response:***

***Yes, the function is linear because the cost of each ticket is the same no matter how many are purchased. For example, $3$ tickets cost $\$18.75$, or $\$6.25$ each. No matter how many tickets are purchased, the cost is $\$6.25$ per ticket.***

b. Describe the limitations of $x$ and $y$.

***The input is a specific number of tickets, so it doesn't make sense for that number to be negative or fractional. The inputs ($x$-values) must be positive integers. The output is the cost, which is okay to be fractional but not negative. The outputs ($y$-values) must be positive rational numbers.***

I need to think about what kinds of inputs would make sense in this situation. For example, does it make sense for the values of $x$ and $y$ to be negative or fractional?

c. Is the function discrete or continuous?

***The function is discrete because you cannot purchase part of a ticket. That is, there is no output that would correspond to $5.25$ tickets.***

I need to use my answer from part (b) for this. Continuous rates can be measured for any input of $x$. Discrete rates are separate and distinct and cannot include fractional parts of an input.

d. Is it reasonable to assume that this function could be used to predict the cost of purchasing $10$ billion tickets? Explain.

***Yes, the function can predict the cost of purchasing $10$ billion tickets. However, it is unlikely that a football stadium could be large enough to hold $10$ billion people.***

2. A function has the table of values below. Examine the information in the table to answer the questions that follow.

| Input | Output |
|---|---|
| 8:00 a.m. | Breakfast |
| 10:00 a.m. | Snack |
| 12:00 p.m. | Lunch |
| 3:00 p.m. | Snack |
| 6:00 p.m. | Dinner |

a. Describe the function.

***It appears that the function describes what kind of meal may be eaten at a particular time of day.***

b. What output would the function assign to the input 8:15 a.m.?

***The function would probably assign breakfast to the input of 8:15 a.m.***

c. Can this function be described using a mathematical rule? Explain.

***Sample Student Response:***

***No, a mathematical rule cannot describe this function. It can be described in words, but there is no formula or rule that can be written.***

I remember my teacher saying that some functions can only be described in words, not numbers or equations, like the problem about fruit and the color of its skin that we did in class.

1. The costs of purchasing certain volumes of gasoline are shown below. We can assume that there is a linear relationship between $x$, the number of gallons purchased, and $y$, the cost of purchasing that many gallons.

| **Number of gallons ($x$)** | 5.4 | 6 | 15 | 17 |
|---|---|---|---|---|
| **Total cost in dollars ($y$)** | 19.71 | 21.90 | 54.75 | 62.05 |

   a. Write an equation that describes $y$ as a linear function of $x$.
   b. Are there any restrictions on the values $x$ and $y$ can adopt?
   c. Is the function discrete?
   d. What number does the linear function assign to 20? Explain what your answer means.

2. A function has the table of values below. Examine the information in the table to answer the questions below.

| **Input** | **Output** |
|---|---|
| one | 3 |
| two | 3 |
| three | 5 |
| four | 4 |
| five | 4 |
| six | 3 |
| seven | 5 |

   a. Describe the function.
   b. What number would the function assign to the word *eleven*?

3. The table shows the distances covered over certain counts of hours traveled by a driver driving a car at a constant speed.

| **Number of hours driven ($x$)** | 3 | 4 | 5 | 6 |
|---|---|---|---|---|
| **Total miles driven ($y$)** | 141 | 188 | 235 | 282 |

   a. Write an equation that describes $y$, the number of miles covered, as a linear function of $x$, number of hours driven.
   b. Are there any restrictions on the value $x$ and $y$ can adopt?
   c. Is the function discrete?
   d. What number does the function assign to 8? Explain what your answer means.
   e. Use the function to determine how much time it would take to drive 500 miles.

4. Consider the function that assigns to each time of a particular day the air temperature at a specific location in Ithaca, NY. The following table shows the values of this function at some specific times.

| | |
|---|---|
| 12:00 noon | 92°F |
| 1:00 p.m. | 90.5°F |
| 2:00 p.m. | 89°F |
| 4:00 p.m. | 86°F |
| 8:00 p.m. | 80°F |

a. Let $y$ represent the air temperature at time $x$ hours past noon. Verify that the data in the table satisfies the linear equation $y = 92 - 1.5x$.

b. Are there any restrictions on the types of values $x$ and $y$ can adopt?

c. Is the function discrete?

d. According to the linear function of part (a), what will the air temperature be at 5:30 p.m.?

e. Is it reasonable to assume that this linear function could be used to predict the temperature for 10:00 a.m. the following day or a temperature at any time on a day next week? Give specific examples in your explanation.

## Exploratory Challenge/Exercises 1–3

1. The distance that Giselle can run is a function of the amount of time she spends running. Giselle runs $3$ miles in $21$ minutes. Assume she runs at a constant rate.

   a. Write an equation in two variables that represents her distance run, $y$, as a function of the time, $x$, she spends running.

   b. Use the equation you wrote in part (a) to determine how many miles Giselle can run in $14$ minutes.

   c. Use the equation you wrote in part (a) to determine how many miles Giselle can run in $28$ minutes.

   d. Use the equation you wrote in part (a) to determine how many miles Giselle can run in $7$ minutes.

e. For a given input $x$ of the function, a time, the matching output of the function, $y$, is the distance Giselle ran in that time. Write the inputs and outputs from parts (b)–(d) as ordered pairs, and plot them as points on a coordinate plane.

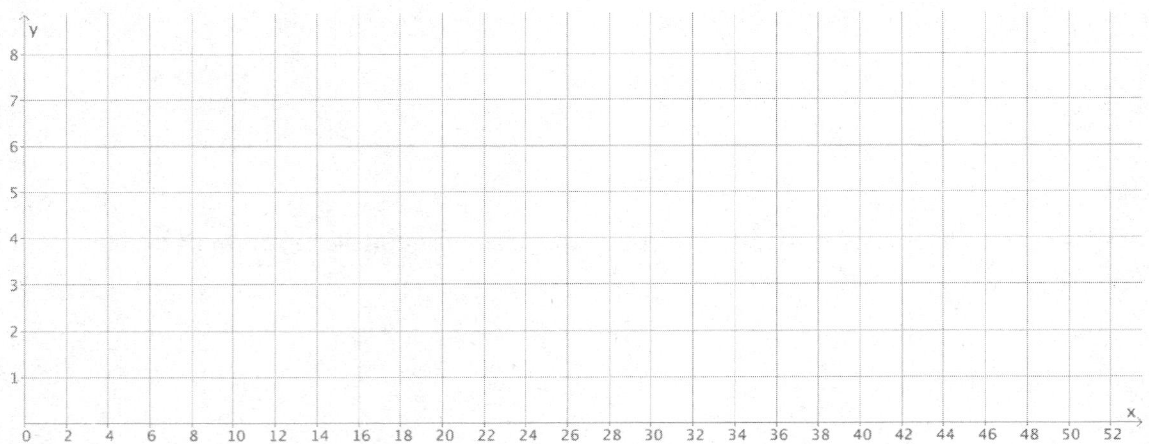

f. What do you notice about the points you plotted?

g. Is the function discrete?

h. Use the equation you wrote in part (a) to determine how many miles Giselle can run in 36 minutes. Write your answer as an ordered pair, as you did in part (e), and include the point on the graph. Is the point in a place where you expected it to be? Explain.

i. Assume you used the rule that describes the function to determine how many miles Giselle can run for any given time and wrote each answer as an ordered pair. Where do you think these points would appear on the graph?

j. What do you think the graph of all the input/output pairs would look like? Explain.

k. Connect the points you have graphed to make a line. Select a point on the graph that has integer coordinates. Verify that this point has an output that the function would assign to the input.

l. Sketch the graph of the equation $y = \frac{1}{7}x$ using the same coordinate plane in part (e). What do you notice about the graph of all the input/output pairs that describes Giselle's constant rate of running and the graph of the equation $y = \frac{1}{7}x$?

2. Sketch the graph of the equation $y = x^2$ for positive values of $x$. Organize your work using the table below, and then answer the questions that follow.

| $x$ | $y$ |
|---|---|
| 0 | |
| 1 | |
| 2 | |
| 3 | |
| 4 | |
| 5 | |
| 6 | |

a. Plot the ordered pairs on the coordinate plane.

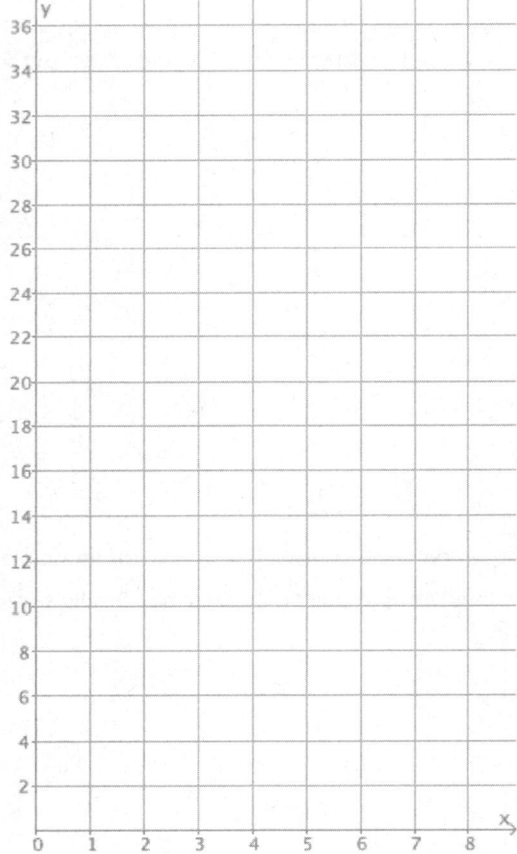

b. What shape does the graph of the points appear to take?

c. Is this equation a linear equation? Explain.

d. Consider the function that assigns to each square of side length $s$ units its area $A$ square units. Write an equation that describes this function.

e. What do you think the graph of all the input/output pairs $(s, A)$ of this function will look like? Explain.

f. Use the function you wrote in part (d) to determine the area of a square with side length 2.5 units. Write the input and output as an ordered pair. Does this point appear to belong to the graph of $y = x^2$?

3. The number of devices a particular manufacturing company can produce is a function of the number of hours spent making the devices. On average, $4$ devices are produced each hour. Assume that devices are produced at a constant rate.

   a. Write an equation in two variables that describes the number of devices, $y$, as a function of the time the company spends making the devices, $x$.

   b. Use the equation you wrote in part (a) to determine how many devices are produced in $8$ hours.

   c. Use the equation you wrote in part (a) to determine how many devices are produced in $6$ hours.

   d. Use the equation you wrote in part (a) to determine how many devices are produced in $4$ hours.

e. The input of the function, $x$, is time, and the output of the function, $y$, is the number of devices produced. Write the inputs and outputs from parts (b)–(d) as ordered pairs, and plot them as points on a coordinate plane.

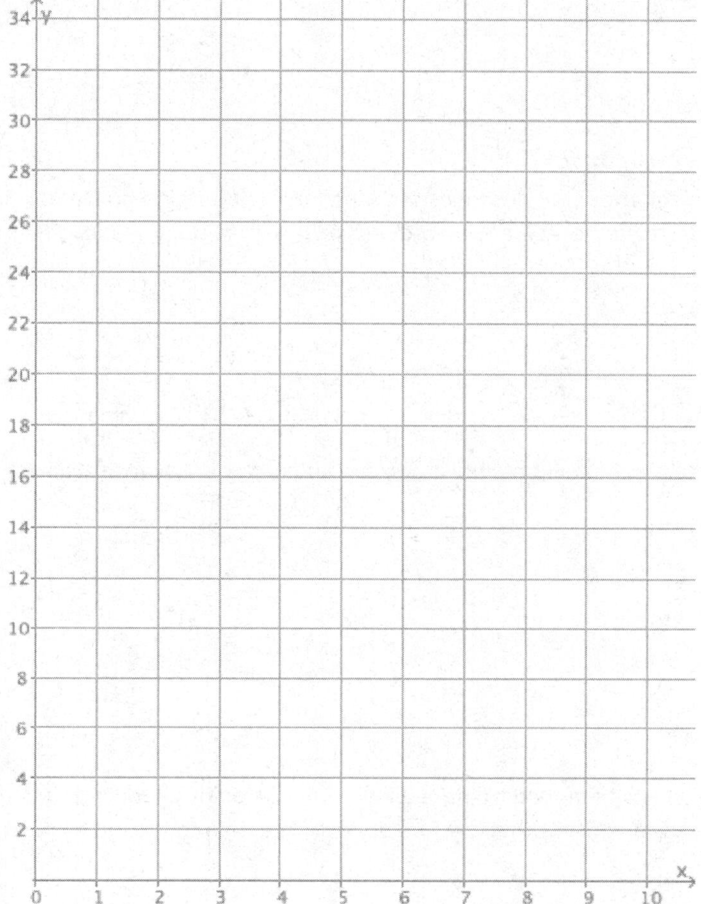


f. What shape does the graph of the points appear to take?

g. Is the function discrete?

h. Use the equation you wrote in part (a) to determine how many devices are produced in $1.5$ hours. Write your answer as an ordered pair, as you did in part (e), and include the point on the graph. Is the point in a place where you expected it to be? Explain.

i. Assume you used the equation that describes the function to determine how many devices are produced for any given time and wrote each answer as an ordered pair. Where do you think these points would appear on the graph?

j. What do you think the graph of all possible input/output pairs will look like? Explain.

k. Connect the points you have graphed to make a line. Select a point on the graph that has integer coordinates. Verify that this point has an output that the function would assign to the input.

l. Sketch the graph of the equation $y = 4x$ using the same coordinate plane in part (e). What do you notice about the graph of the input/output pairs that describes the company's constant rate of producing devices and the graph of the equation $y = 4x$?

## Exploratory Challenge/Exercise 4

4. Examine the three graphs below. Which, if any, could represent the graph of a function? Explain why or why not for each graph.

Graph 1:

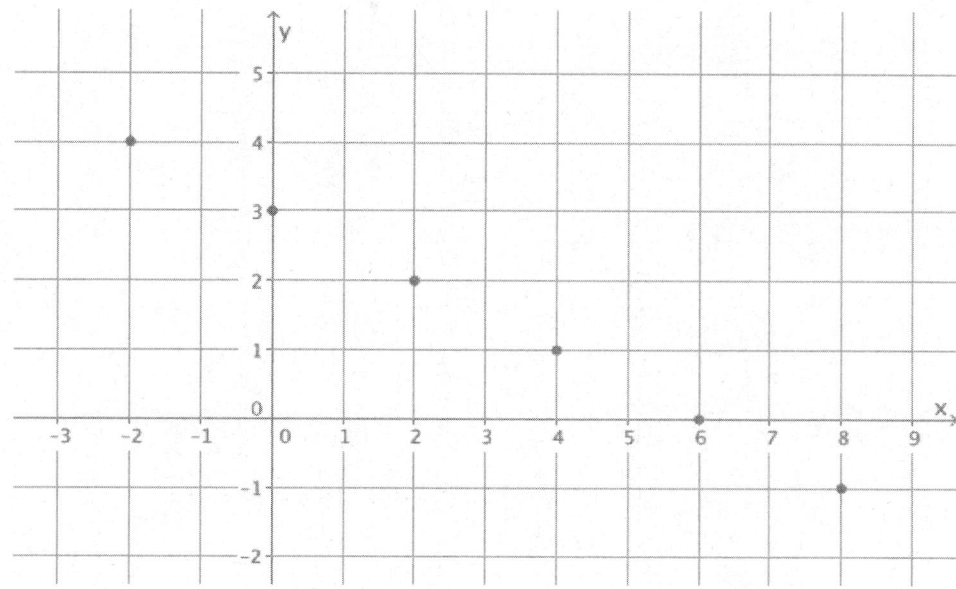

Graph 2:

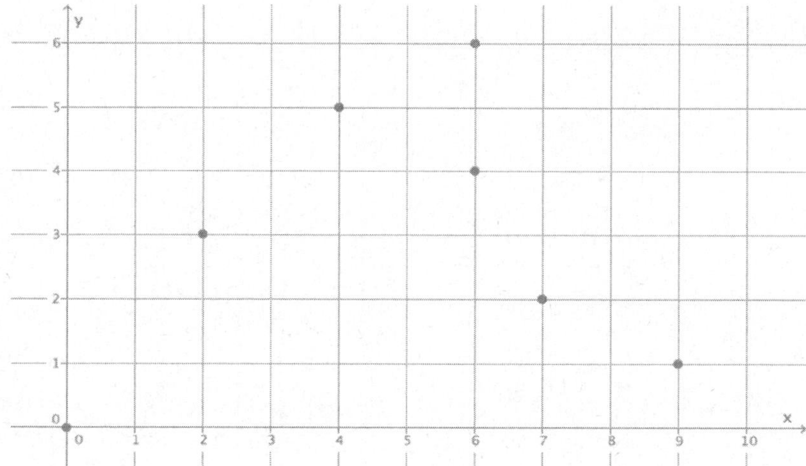

Graph 3:

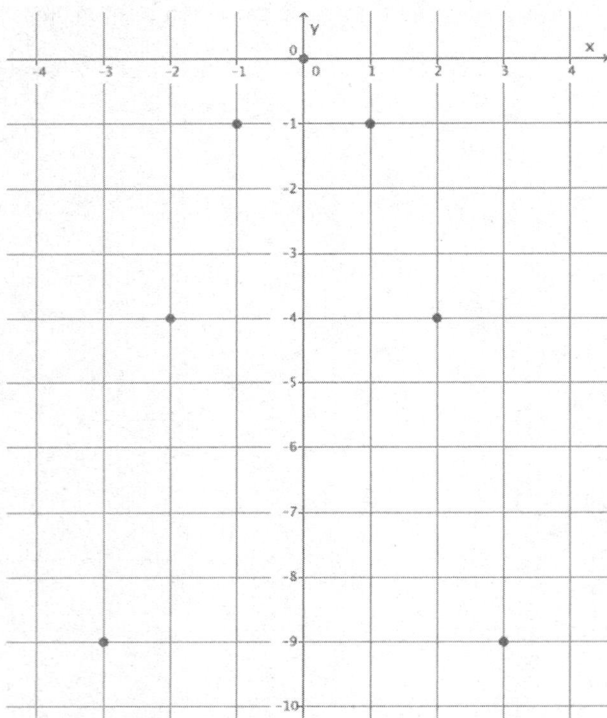

### Lesson Summary

The graph of a function is defined to be the set of all points $(x, y)$ with $x$ an input for the function and $y$ its matching output.

If a function can be described by an equation, then the graph of the function is the same as the graph of the equation that represents it (at least at points which correspond to valid inputs of the function).

It is not possible for two different points in the plot of the graph of a function to have the same $x$-coordinate.

Name ____________________________________ Date ________________

Water flows from a hose at a constant rate of 11 gallons every 4 minutes. The total amount of water that flows from the hose is a function of the number of minutes you are observing the hose.

a. Write an equation in two variables that describes the amount of water, $y$, in gallons, that flows from the hose as a function of the number of minutes, $x$, you observe it.

b. Use the equation you wrote in part (a) to determine the amount of water that flows from the hose during an 8-minute period, a 4-minute period, and a 2-minute period.

c. An input of the function, $x$, is time in minutes, and the output of the function, $y$, is the amount of water that flows out of the hose in gallons. Write the inputs and outputs from part (b) as ordered pairs, and plot them as points on the coordinate plane.

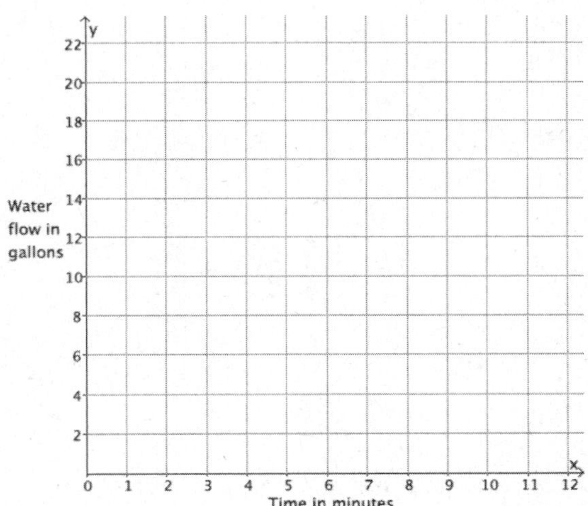

1. Graph the equation $y = x^2 + 2$ for positive values of $x$. Organize your work using the table below, and then answer the questions that follow.

| $x$ | $y$ |
|---|---|
| 0 | $\mathbf{0^2 + 2 = 2}$ |
| 1 | $\mathbf{1^2 + 2 = 3}$ |
| 2 | $\mathbf{2^2 + 2 = 6}$ |
| 3 | $\mathbf{3^2 + 2 = 11}$ |
| 4 | $\mathbf{4^2 + 2 = 18}$ |

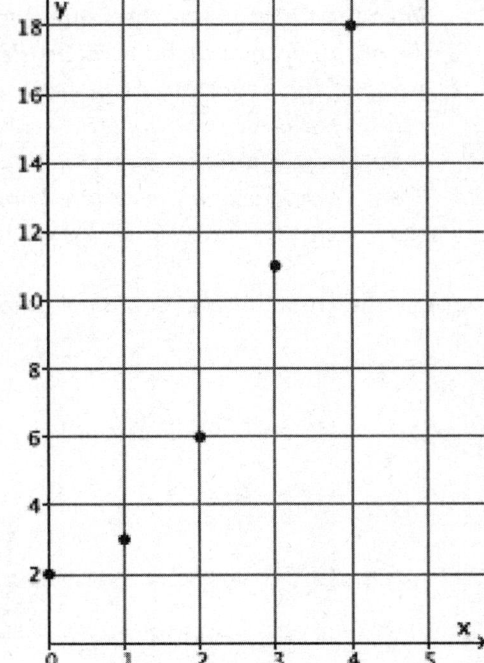


a. Plot the ordered pairs on the coordinate plane.

***The ordered pairs are*** $\mathbf{(0, 2)}$, $\mathbf{(1, 3)}$, $\mathbf{(2, 6)}$, $\mathbf{(3, 11)}$ ***and*** $\mathbf{(4, 18)}$.

b. What shape does the graph of the points appear to take?

***It appears to be a curve.***

Does the graph appear to take the shape of a line or a curve? Can I draw one straight line using my straightedge through all of the points?

c. Is this the graph of a linear equation? Explain.

***No. A graph that is linear would have the shape of a line. This graph is a curve.***

d. A function has the rule so that it assigns to each input, $x$, the output, $x^2 + 2$. The rule for this function is $y = x^2 + 2$. What do you think the graph of this function will look like? Explain.

***Since the function has the same rule as the equation, the graph of the function will be identical to the graph of the equation. I can verify this by taking each input, $x$, and substituting it into the equation that describes the function $y = x^2 + 2$ to get the output. Then, I would graph the ordered pairs (input, output).***

The equation $y = x^2 + 2$ and the rule for this function are the same. What will that mean about their graphs?

2. Examine the graph below. Could the graph represent the graph of a function? Explain why or why not.

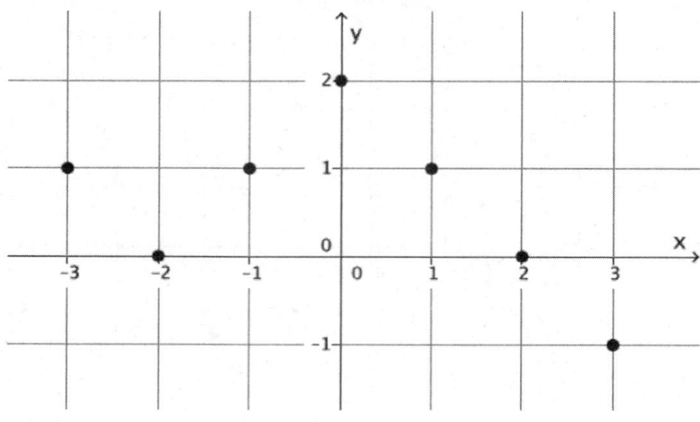


The definition of a function will help me. I need to check that every input ($x$-value) has only one output ($y$-value).

***This is the graph of a function because every $x$-value has just one corresponding $y$-value.***

3. Examine the graph below. Could the graph represent the graph of a function? Explain why or why not.

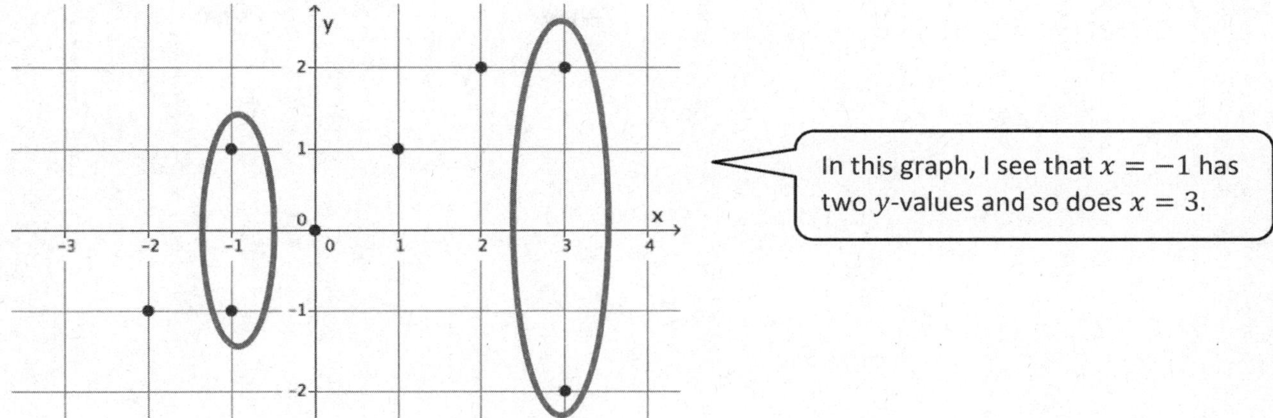


***This is <u>not</u> the graph of a function because there are some values of $x$ (inputs) that have more than one corresponding $y$-value (outputs). For example, the input of $-1$ corresponds to both $1$ and $-1$. Another example is the input of $3$; it corresponds to the outputs of $2$ and $-2$.***

1. The distance that Scott walks is a function of the time he spends walking. Scott can walk $\frac{1}{2}$ mile every 8 minutes. Assume he walks at a constant rate.
   a. Predict the shape of the graph of the function. Explain.
   b. Write an equation to represent the distance that Scott can walk in miles, $y$, in $x$ minutes.
   c. Use the equation you wrote in part (b) to determine how many miles Scott can walk in 24 minutes.
   d. Use the equation you wrote in part (b) to determine how many miles Scott can walk in 12 minutes.
   e. Use the equation you wrote in part (b) to determine how many miles Scott can walk in 16 minutes.
   f. Write your inputs and corresponding outputs as ordered pairs, and then plot them on a coordinate plane.

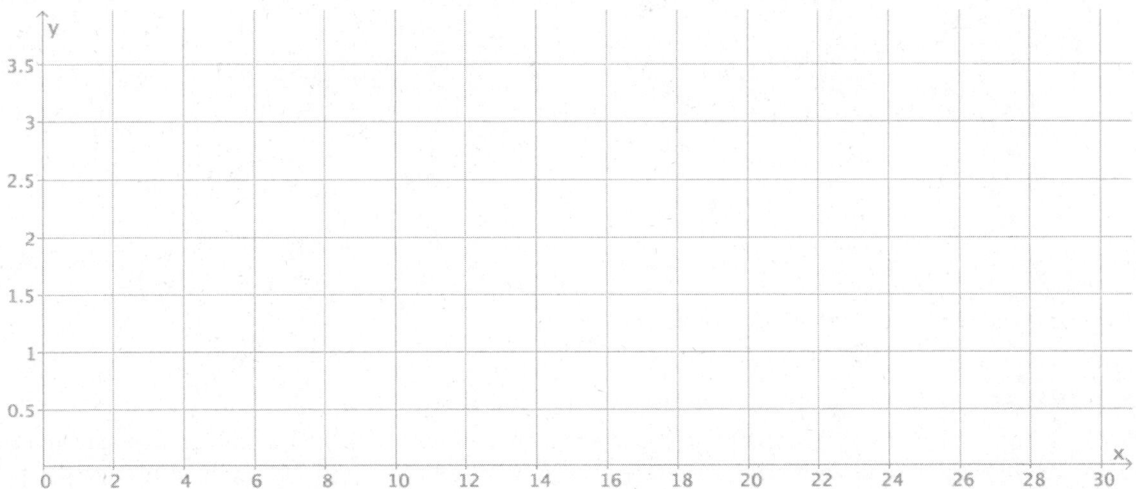


   g. What shape does the graph of the points appear to take? Does it match your prediction?
   h. Connect the points to make a line. What is the equation of the line?

2. Graph the equation $y = x^3$ for positive values of $x$. Organize your work using the table below, and then answer the questions that follow.

| $x$ | $y$ |
|---|---|
| 0 | |
| 0.5 | |
| 1 | |
| 1.5 | |
| 2 | |
| 2.5 | |

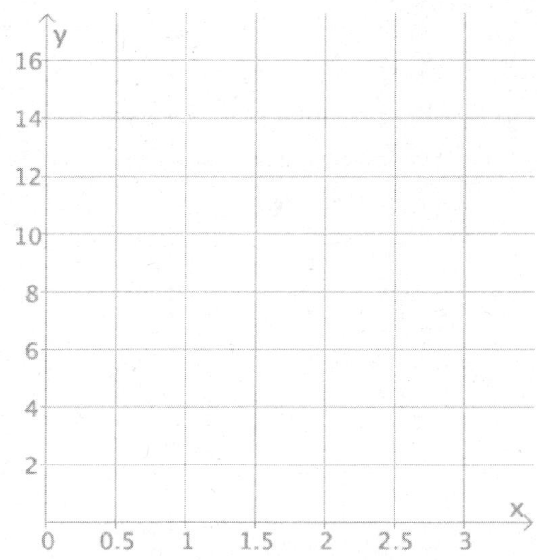


a. Plot the ordered pairs on the coordinate plane.

b. What shape does the graph of the points appear to take?

c. Is this the graph of a linear function? Explain.

d. Consider the function that assigns to each positive real number $s$ the volume $V$ of a cube with side length $s$ units. An equation that describes this function is $V = s^3$. What do you think the graph of this function will look like? Explain.

e. Use the function in part (d) to determine the volume of a cube with side length of 3 units. Write the input and output as an ordered pair. Does this point appear to belong to the graph of $y = x^3$?

3. Sketch the graph of the equation $y = 180(x - 2)$ for whole numbers. Organize your work using the table below, and then answer the questions that follow.

| $x$ | $y$ |
|---|---|
| 3 | |
| 4 | |
| 5 | |
| 6 | |

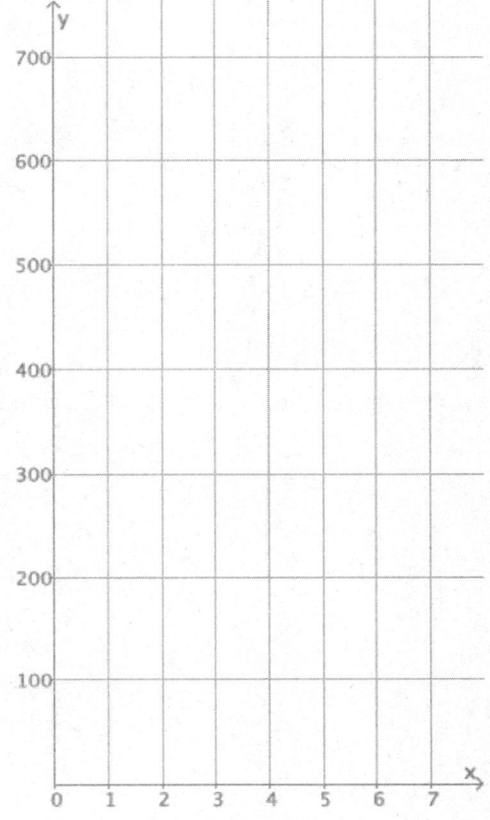


a. Plot the ordered pairs on the coordinate plane.

b. What shape does the graph of the points appear to take?

c. Is this graph a graph of a function? How do you know?

d. Is this a linear equation? Explain.

e. The sum $S$ of interior angles, in degrees, of a polygon with $n$ sides is given by $S = 180(n - 2)$. If we take this equation as defining $S$ as a function of $n$, how do think the graph of this $S$ will appear? Explain.

f. Is this function discrete? Explain.

4. Examine the graph below. Could the graph represent the graph of a function? Explain why or why not.

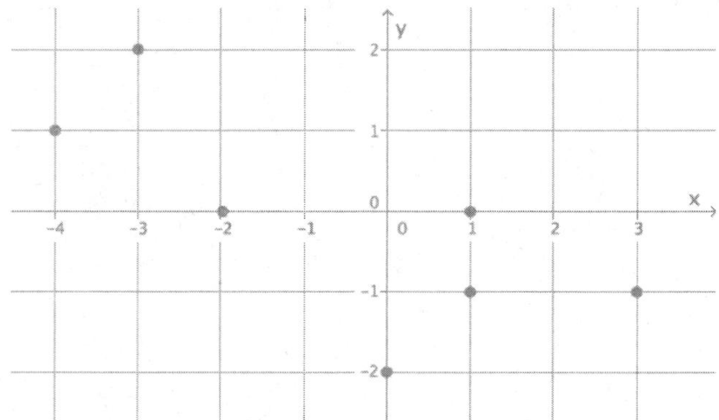

5. Examine the graph below. Could the graph represent the graph of a function? Explain why or why not.

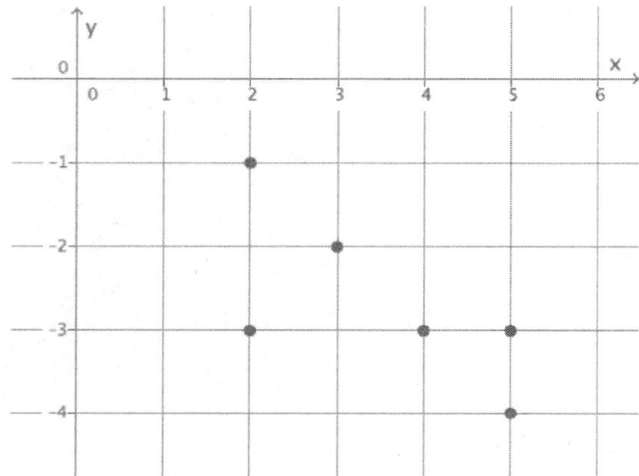

6. Examine the graph below. Could the graph represent the graph of a function? Explain why or why not.

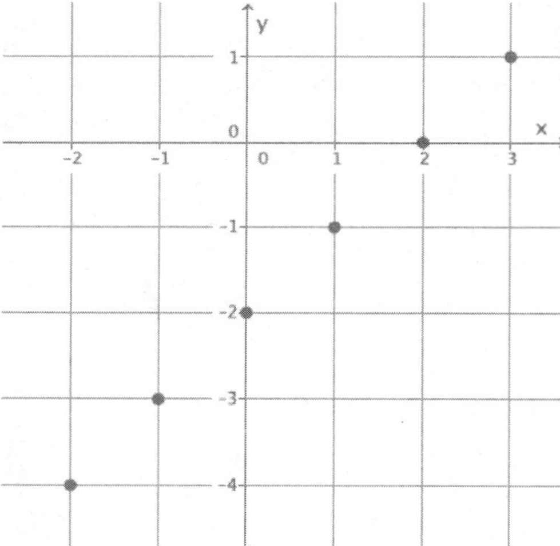

## Opening Exercise

A function is said to be linear if the rule defining the function can be described by a linear equation.

Functions 1, 2, and 3 have table-values as shown. Which of these functions appear to be linear? Justify your answers.

| Input | Output |
|---|---|
| 2 | 5 |
| 4 | 7 |
| 5 | 8 |
| 8 | 11 |

| Input | Output |
|---|---|
| 2 | 4 |
| 3 | 9 |
| 4 | 16 |
| 5 | 25 |

| Input | Output |
|---|---|
| 0 | −3 |
| 1 | 1 |
| 2 | 6 |
| 3 | 9 |

## Exercise

A function assigns the inputs shown the corresponding outputs given in the table below.

| Input | Output |
|---|---|
| 1 | 2 |
| 2 | −1 |
| 4 | −7 |
| 6 | −13 |

a. Do you suspect the function is linear? Compute the rate of change of this data for at least three pairs of inputs and their corresponding outputs.

b. What equation seems to describe the function?

c. As you did not verify that the rate of change is constant across <u>all</u> input/output pairs, check that the equation you found in part (a) does indeed produce the correct output for each of the four inputs 1, 2, 4, and 6.

d. What will the graph of the function look like? Explain.

**Lesson Summary**

If the rate of change for pairs of inputs and corresponding outputs for a function is the same for all pairs (constant), then the function is a linear function. It can thus be described by a linear equation $y = mx + b$.

The graph of a linear function will be a set of points contained in a line. If the linear function is discrete, then its graph will be a set of distinct collinear points. If the linear function is not discrete, then its graph will be a full straight line or a portion of the line (as appropriate for the context of the problem).

Name ____________________________________________ Date ____________________

1. Sylvie claims that a function with the table of inputs and outputs below is a linear function. Is she correct? Explain.

| Input | Output |
|---|---|
| $-3$ | $-25$ |
| $2$ | $10$ |
| $5$ | $31$ |
| $8$ | $54$ |

2. A function assigns the inputs and corresponding outputs shown in the table to the right.

| Input | Output |
|---|---|
| $-2$ | $3$ |
| $8$ | $-2$ |
| $10$ | $-3$ |
| $20$ | $-8$ |

   a. Does the function appear to be linear? Check at least three pairs of inputs and their corresponding outputs.

b. Can you write a linear equation that describes the function?

c. What will the graph of the function look like? Explain.

1. A function assigns to the inputs the corresponding outputs shown in the table below.

| Input ($x$) | Output ($y$) |
|---|---|
| −3 | 5 |
| −1 | 7 |
| 1 | 9 |
| 3 | 11 |

a. Is the function a linear function? Check at least three pairs of inputs and their corresponding outputs.

$$\frac{7-5}{-1-(-3)}=\frac{2}{2}=1 \quad \frac{11-7}{3-(-1)}=\frac{4}{4}=1 \quad \frac{11-9}{3-1}=\frac{2}{2}=1$$

I need to make sure the rate of change is the same value for each of the three pairs that I check. If they are the same, then I know the function is a linear function.

b. What equation describes the function?

***Using the input and corresponding output*** $(1, 9)$:

$$y = mx + b$$
$$9 = 1(1) + b$$
$$9 = 1 + b$$
$$9 - 1 = 1 - 1 + b$$
$$8 = b$$

***Since*** $m = 1$ ***and*** $b = 8$***, the equation that describes this function is*** $y = 1x + 8$ ***or just*** $y = x + 8$.

c. What will the graph of the function look like? Explain.

***Since the function is described by a linear function, and I know from the last lesson that the graph of the function will be identical to the graph of the equation that describes it, then the graph of this function is a line. Linear equations graph as lines; therefore, linear functions will also graph as lines.***

2. Is the following graph a graph of a linear function? How would you determine if it is a linear function?

> This is just like part (a) of the last problem. The only difference is that my inputs and outputs are graphed instead of in a table.

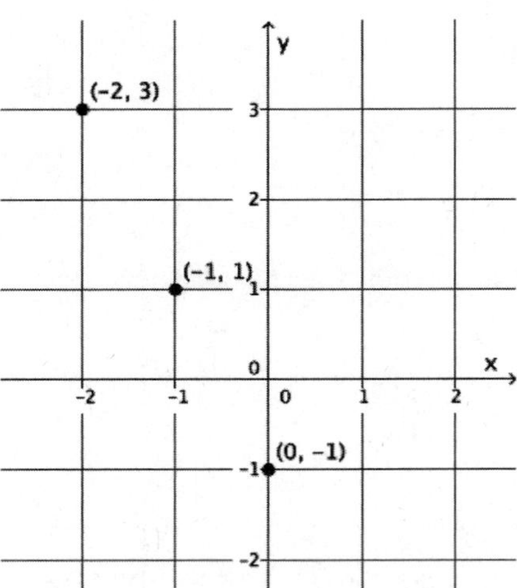


***If the rate of change is the same, then this is a linear function.***

$$\frac{3-1}{-2-(-1)} = \frac{2}{-1} = -2$$

$$\frac{-3-(-1)}{1-0} = \frac{-2}{1} = -2$$

$$\frac{-1-1}{0-(-1)} = \frac{-2}{1} = -2$$

***Since the rate of change is the same, −2, this is the graph of a linear function.***

3. Xander says you really only need to check two pairs of inputs and outputs to determine if the function is linear. Is he correct? Explain. Hint: Show an example with a table where this is not true.

***It is always a good idea to check three pairs of inputs and outputs.***

***The table below demonstrates why.***

> I need to develop a table of values where two pairs of inputs and outputs give the same value, but a third pair would give a different value.

| *Input* ($x$) | *Output* ($y$) |
|---|---|
| 3 | 7 |
| −1 | −1 |
| 0 | 1 |
| 5 | 6 |

$$\frac{7-(-1)}{3-(-1)} = \frac{8}{4} = 2 \qquad \frac{1-(-1)}{0-(-1)} = \frac{2}{1} = 2 \qquad \frac{6-1}{5-0} = \frac{5}{5} = 1$$

***Since the third pair gave a different value than the first two pairs, it shows that Xander's statement is incorrect.***

1. A function assigns to the inputs given the corresponding outputs shown in the table below.

| Input | Output |
|---|---|
| 3 | 9 |
| 9 | 17 |
| 12 | 21 |
| 15 | 25 |

 a. Does the function appear to be linear? Check at least three pairs of inputs and their corresponding outputs.
 b. Find a linear equation that describes the function.
 c. What will the graph of the function look like? Explain.

2. A function assigns to the inputs given the corresponding outputs shown in the table below.

| Input | Output |
|---|---|
| $-1$ | 2 |
| 0 | 0 |
| 1 | 2 |
| 2 | 8 |
| 3 | 18 |

 a. Is the function a linear function?
 b. What equation describes the function?

3. A function assigns the inputs and corresponding outputs shown in the table below.

| Input | Output |
|---|---|
| 0.2 | 2 |
| 0.6 | 6 |
| 1.5 | 15 |
| 2.1 | 21 |

 a. Does the function appear to be linear? Check at least three pairs of inputs and their corresponding outputs.
 b. Find a linear equation that describes the function.
 c. What will the graph of the function look like? Explain.

4. Martin says that you only need to check the first and last input and output values to determine if the function is linear. Is he correct? Explain.

5. Is the following graph a graph of a linear function? How would you determine if it is a linear function?

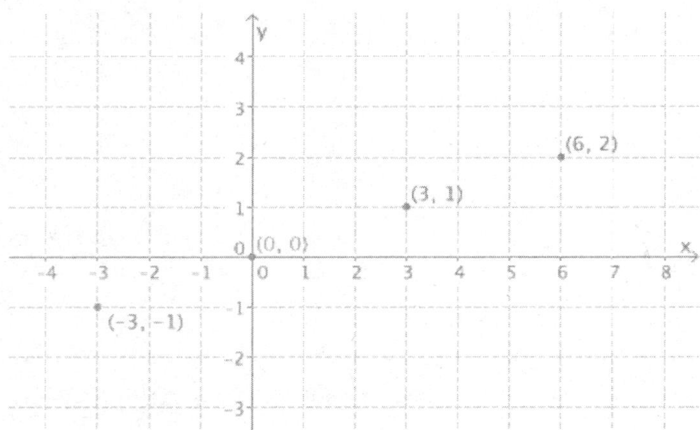


6. A function assigns to the inputs given the corresponding outputs shown in the table below.

| Input | Output |
|---|---|
| $-6$ | $-6$ |
| $-5$ | $-5$ |
| $-4$ | $-4$ |
| $-2$ | $-2$ |

a. Does the function appear to be a linear function?

b. What equation describes the function?

c. What will the graph of the function look like? Explain.

## Exploratory Challenge/Exercises 1–4

Each of Exercises 1–4 provides information about two functions. Use that information given to help you compare the two functions and answer the questions about them.

1. Alan and Margot each drive from City A to City B, a distance of $147$ miles. They take the same route and drive at constant speeds. Alan begins driving at 1:40 p.m. and arrives at City B at 4:15 p.m. Margot's trip from City A to City B can be described with the equation $y = 64x$, where $y$ is the distance traveled in miles and $x$ is the time in minutes spent traveling. Who gets from City A to City B faster?

2. You have recently begun researching phone billing plans. Phone Company A charges a flat rate of $75 a month. A flat rate means that your bill will be $75 each month with no additional costs. The billing plan for Phone Company B is a linear function of the number of texts that you send that month. That is, the total cost of the bill changes each month depending on how many texts you send. The table below represents some inputs and the corresponding outputs that the function assigns.

| **Input (number of texts)** | **Output (cost of bill in dollars)** |
|---|---|
| $50$ | $50$ |
| $150$ | $60$ |
| $200$ | $65$ |
| $500$ | $95$ |

At what number of texts would the bill from each phone plan be the same? At what number of texts is Phone Company A the better choice? At what number of texts is Phone Company B the better choice?

3. The function that gives the volume of water, $y$, that flows from Faucet A in gallons during $x$ minutes is a linear function with the graph shown. Faucet B's water flow can be described by the equation $y = \frac{5}{6}x$, where $y$ is the volume of water in gallons that flows from the faucet during $x$ minutes. Assume the flow of water from each faucet is constant. Which faucet has a faster rate of flow of water? Each faucet is being used to fill a tub with a volume of 50 gallons. How long will it take each faucet to fill its tub? How do you know?

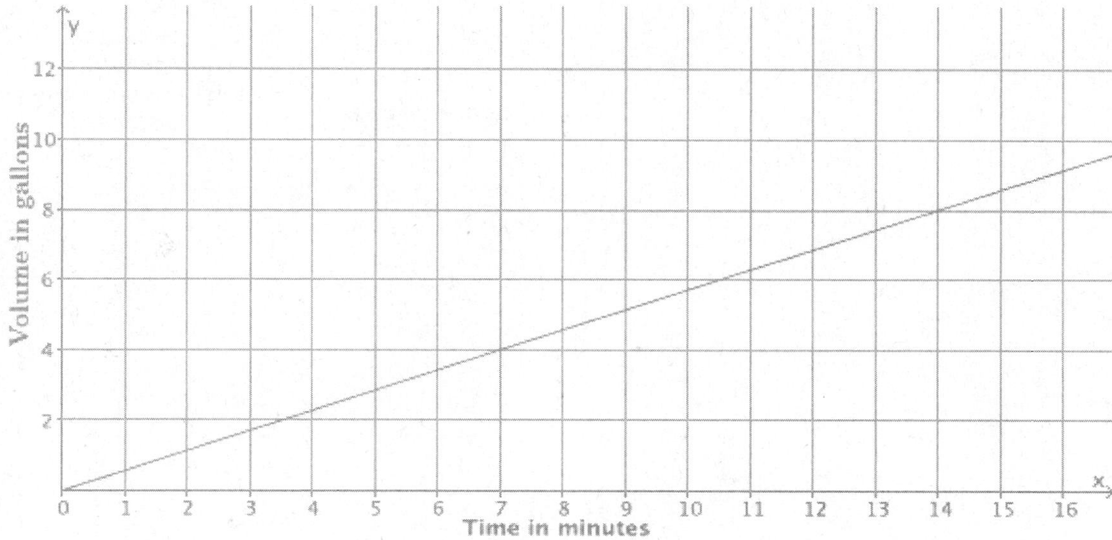


Suppose the tub being filled by Faucet A already had 15 gallons of water in it, and the tub being filled by Faucet B started empty. If now both faucets are turned on at the same time, which faucet will fill its tub fastest?

4. Two people, Adam and Bianca, are competing to see who can save the most money in one month. Use the table and the graph below to determine who will save the most money at the end of the month. State how much money each person had at the start of the competition. (Assume each is following a linear function in his or her saving habit.)

Adam's Savings:

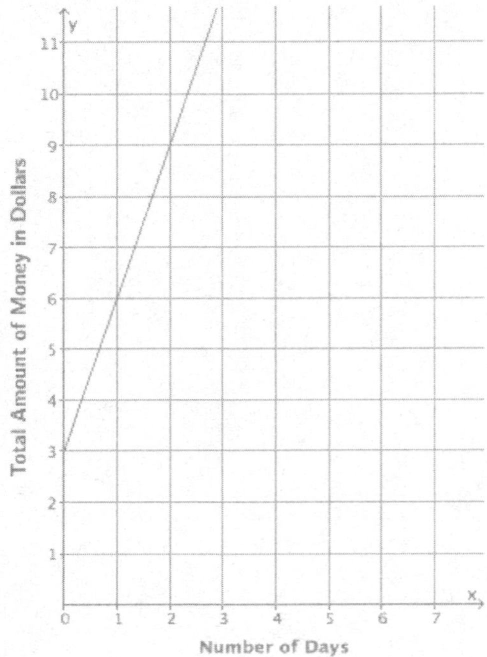


Bianca's Savings:

| **Input (Number of Days)** | **Output (Total amount of money in dollars)** |
|---|---|
| 5 | 17 |
| 8 | 26 |
| 12 | 38 |
| 20 | 62 |

Name ____________________________ Date ______________

Brothers Paul and Pete walk 2 miles to school from home. Paul can walk to school in 24 minutes. Pete has slept in again and needs to run to school. Paul walks at a constant rate, and Pete runs at a constant rate. The graph of the function that represents Pete's run is shown below.

a. Which brother is moving at a greater rate? Explain how you know.

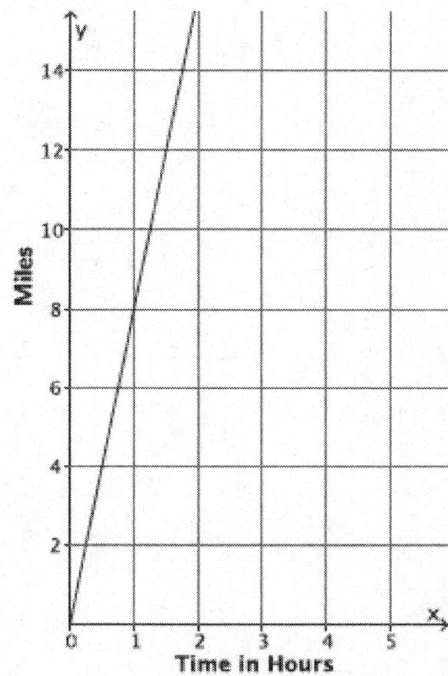


b. If Pete leaves 5 minutes after Paul, will he catch up to Paul before they get to school?

1. The graph below represents the distance, $y$, Car A travels in $x$ minutes. The table represents the distance, $y$, Car B travels in $x$ minutes.

Car A:

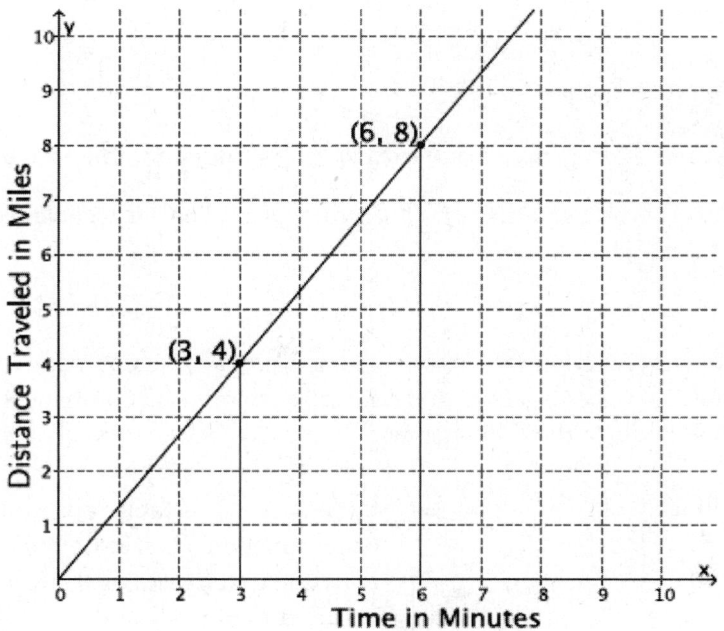


If this car is traveling at a constant rate, I can compute its rate by using the slope formula or looking at the fraction that compares the vertical distance to the horizontal distance between any pair of points.

Car B:

| Time in Minutes ($x$) | Distance in Miles ($y$) |
|---|---|
| 15 | 13.5 |
| 25 | 22.5 |
| 35 | 31.5 |

To figure out if this car is traveling at a constant rate, I need to check various pairs of data using the slope formula to see if they are equal to the same value. That value, if constant, will be the rate the car travels.

a. Is Car A traveling at a constant rate? Explain how you know.

***Since the graph of the data related to Car A is a line, the equation that describes the function must be a linear equation. Therefore, this car is traveling at a constant rate.***

b. Is Car B traveling at a constant rate? Explain how you know.

$\frac{22.5-13.5}{25-15}=\frac{9}{10}$ $\frac{31.5-22.5}{35-25}=\frac{9}{10}$ $\frac{31.5-13.5}{35-15}=\frac{18}{20}=\frac{9}{10}$

***Car B is traveling at a constant rate because all pairs of data have the same slope value, or rate of change.***

c. Which car is traveling at a slower rate? Explain.

***Using the graph or the slope formula, Car A travels at a rate of*** $\frac{4}{3}$***. By inspecting the rate of change for the data within the table, Car B travels at a rate of*** $\frac{9}{10}$***. Since*** $\frac{9}{10}<\frac{4}{3}$***, Car B is traveling at a slower rate.***

2. The rule $y=6.67x+35$ describes the cost function for a phone plan at Company A. Company A charges a flat fee of $35 for phone service, plus $6.67 per gigabyte of data used each month. Company B has a similar function that assigns the values shown in the table below.

| Gigabytes of Data ($x$) | Total Cost in Dollars ($y$) |
|---|---|
| 1 | 39.50 |
| 3 | 54.50 |
| 5 | 69.50 |

If the data in this table represent a linear function, I can write the equation that describes the function in the form of $y=mx+b$. Then, I can compare the rates for data, $m$, and the flat fee, $b$.

***We need to check that the data in the table represent a linear function.***

$\frac{54.50-39.50}{3-1}=\frac{15}{2}=7.5$ $\frac{69.50-54.50}{5-3}=\frac{15}{2}=7.5$ $\frac{69.50-39.50}{5-1}=\frac{30}{4}=7.5$

***Since the rate of change is equal to the same constant,*** $7.5$***, I can write the equation that describes the cost function for Company B. Using the input,*** $x$***, and output,*** $y$***,*** $(1, 39.50)$***:***

$$y=mx+b$$
$$39.50=7.5(1)+b$$
$$39.50=7.5+b$$
$$39.50-7.50=7.50-7.50+b$$
$$32=b$$

***The equation that describes the cost function for Company B is*** $y=7.5x+32$***.***

a. Which company charges a higher rate for data usage?

***Comparing the rates,*** $7.5>6.67$***, we can conclude that Company B charges a higher rate for data usage.***

b. Which company charges a higher flat fee for phone service?

***Comparing the flat fees,*** $35 > 32$***, we can conclude that Company A charges a higher flat fee.***

c. At what number of gigabytes of data used would both companies charge the same amount of money? How much will the total cost be for that amount of gigabytes used?

$$\begin{cases} y = 6.67x + 35 \\ y = 7.5x + 32 \end{cases}$$

***Since both equations are equal to*** $y$***, I can write the expressions on the right of the equal sign as equal to one another and then solve.***

$$6.67x + 35 = 7.5x + 32$$

$$6.67x - 6.67x + 35 - 32 = 7.5x - 6.67x + 32 - 32$$

$$3 = 0.83x$$

$$\frac{3}{0.83} = x$$

$$3.61 \approx x$$

If the data related to both companies were graphed on the same coordinate plane, then the point of intersection of their lines would be when the costs were equal. I should write and solve a system of equations to answer this question.

***Now, substitute the value of*** $x$ ***into the first equation and solve.***

$$y \approx 6.67(3.61) + 35$$

$$y \approx 59.08$$

***At about*** $3.61$ ***gigabytes, the cost would be the same at both companies. That cost would be about*** $\$59.08$***.***

1. The graph below represents the distance in miles, $y$, Car A travels in $x$ minutes. The table represents the distance in miles, $y$, Car B travels in $x$ minutes. It is moving at a constant rate. Which car is traveling at a greater speed? How do you know?

Car A:

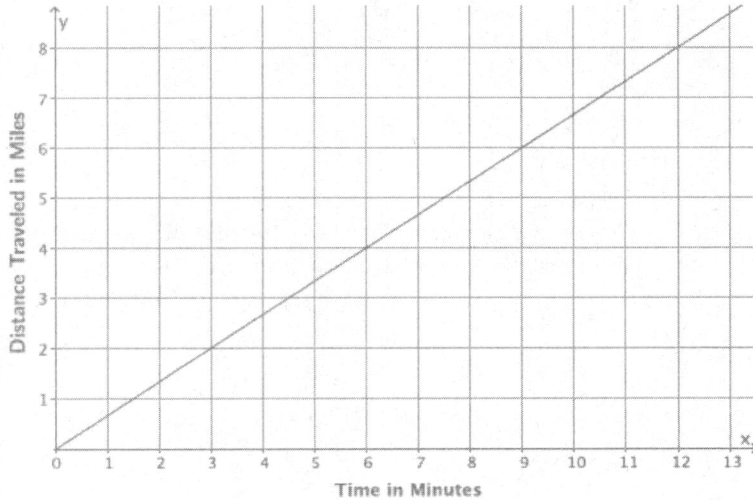


Car B:

| Time in minutes ($x$) | Distance in miles ($y$) |
|---|---|
| 15 | 12.5 |
| 30 | 25 |
| 45 | 37.5 |

2. The local park needs to replace an existing fence that is 6 feet high. Fence Company A charges \$7,000 for building materials and \$200 per foot for the length of the fence. Fence Company B charges are based solely on the length of the fence. That is, the total cost of the six-foot high fence will depend on how long the fence is. The table below represents some inputs and their corresponding outputs that the cost function for Fence Company B assigns. It is a linear function.

| Input (length of fence in feet) | Output (cost of bill in dollars) |
|---|---|
| 100 | 26,000 |
| 120 | 31,200 |
| 180 | 46,800 |
| 250 | 65,000 |

a. Which company charges a higher rate per foot of fencing? How do you know?

b. At what number of the length of the fence would the cost from each fence company be the same? What will the cost be when the companies charge the same amount? If the fence you need were 190 feet in length, which company would be a better choice?

3. The equation $y = 123x$ describes the function for the number of toys, $y$, produced at Toys Plus in $x$ minutes of production time. Another company, #1 Toys, has a similar function, also linear, that assigns the values shown in the table below. Which company produces toys at a slower rate? Explain.

| Time in minutes $(x)$ | Toys Produced $(y)$ |
|---|---|
| 5 | 600 |
| 11 | 1,320 |
| 13 | 1,560 |

4. A train is traveling from City A to City B, a distance of 320 miles. The graph below shows the number of miles, $y$, the train travels as a function of the number of hours, $x$, that have passed on its journey. The train travels at a constant speed for the first four hours of its journey and then slows down to a constant speed of 48 miles per hour for the remainder of its journey.

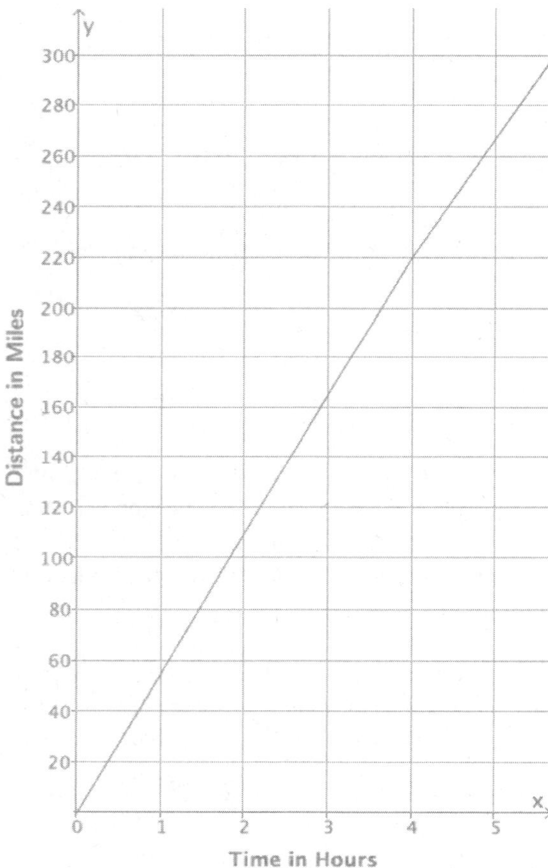


a. How long will it take the train to reach its destination?

b. If the train had not slowed down after 4 hours, how long would it have taken to reach its destination?

c. Suppose after 4 hours, the train increased its constant speed. How fast would the train have to travel to complete the destination in 1.5 hours?

5.

a. A hose is used to fill up a 1,200 gallon water truck. Water flows from the hose at a constant rate. After 10 minutes, there are 65 gallons of water in the truck. After 15 minutes, there are 82 gallons of water in the truck. How long will it take to fill up the water truck? Was the tank initially empty?

b. The driver of the truck realizes that something is wrong with the hose he is using. After 30 minutes, he shuts off the hose and tries a different hose. The second hose flows at a constant rate of 18 gallons per minute. How long now does it take to fill up the truck?

## Exploratory Challenge/Exercises 1–3

1. Consider the function that assigns to each number $x$ the value $x^2$.

   a. Do you think the function is linear or nonlinear? Explain.

   b. Develop a list of inputs and outputs for this function. Organize your work using the table below. Then, answer the questions that follow.

| **Input** ($x$) | **Output** ($x^2$) |
|---|---|
| $-5$ | |
| $-4$ | |
| $-3$ | |
| $-2$ | |
| $-1$ | |
| $0$ | |
| $1$ | |
| $2$ | |
| $3$ | |
| $4$ | |
| $5$ | |

c. Plot the inputs and outputs as ordered pairs defining points on the coordinate plane.

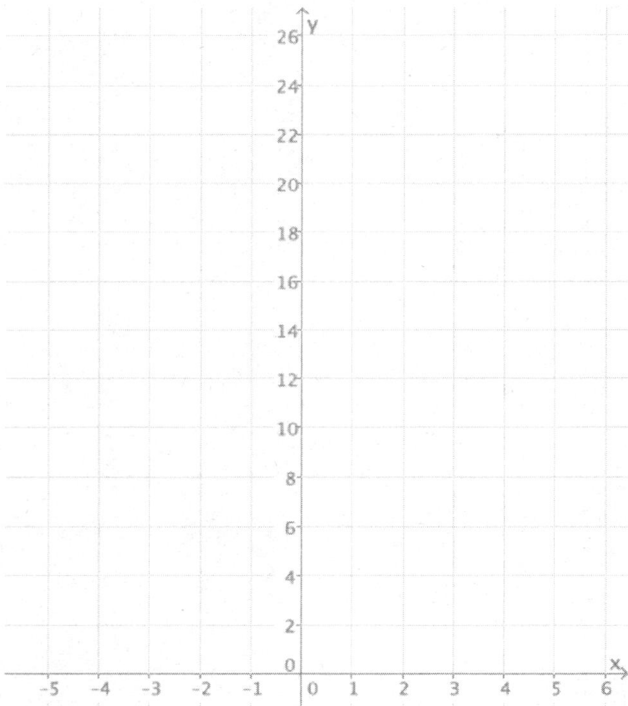


d. What shape does the graph of the points appear to take?

e. Find the rate of change using rows 1 and 2 from the table above.

f. Find the rate of change using rows 2 and 3 from the table above.

g. Find the rate of change using any two other rows from the table above.

h. Return to your initial claim about the function. Is it linear or nonlinear? Justify your answer with as many pieces of evidence as possible.

2. Consider the function that assigns to a number $x$ the value $x^3$.

a. Do you think the function is linear or nonlinear? Explain.

b. Develop a list of inputs and outputs for this function. Organize your work using the table below. Then, answer the questions that follow.

| Input ($x$) | Output ($x^3$) |
|---|---|
| $-2.5$ | |
| $-2$ | |
| $-1.5$ | |
| $-1$ | |
| $-0.5$ | |
| $0$ | |
| $0.5$ | |
| $1$ | |
| $1.5$ | |
| $2$ | |
| $2.5$ | |

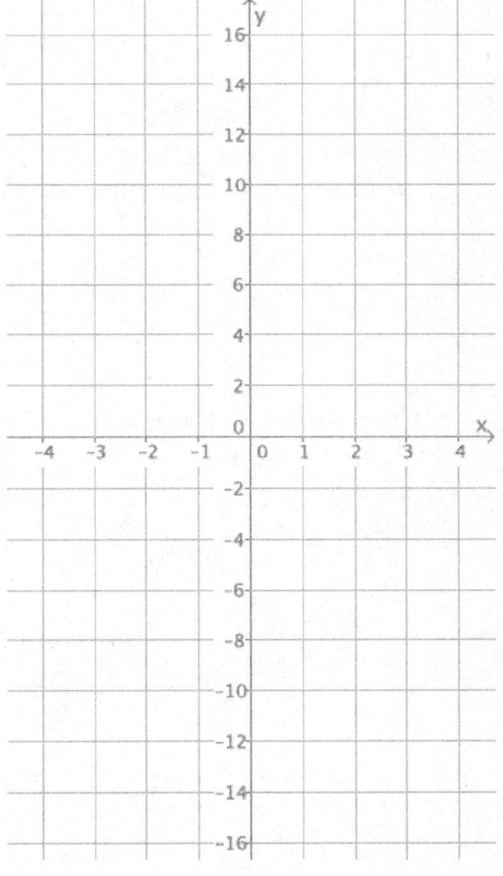

c. Plot the inputs and outputs as ordered pairs defining points on the coordinate plane.

d. What shape does the graph of the points appear to take?

e. Find the rate of change using rows 2 and 3 from the table above.

f. Find the rate of change using rows 3 and 4 from the table above.

g. Find the rate of change using rows 8 and 9 from the table above.

h. Return to your initial claim about the function. Is it linear or nonlinear? Justify your answer with as many pieces of evidence as possible.

3. Consider the function that assigns to each positive number $x$ the value $\frac{1}{x}$.

a. Do you think the function is linear or nonlinear? Explain.

b. Develop a list of inputs and outputs for this function. Organize your work using the table below. Then, answer the questions that follow.

| Input ($x$) | Output $\left(\frac{1}{x}\right)$ |
|---|---|
| 0.1 | |
| 0.2 | |
| 0.4 | |
| 0.5 | |
| 0.8 | |
| 1 | |
| 1.6 | |
| 2 | |
| 2.5 | |
| 4 | |
| 5 | |

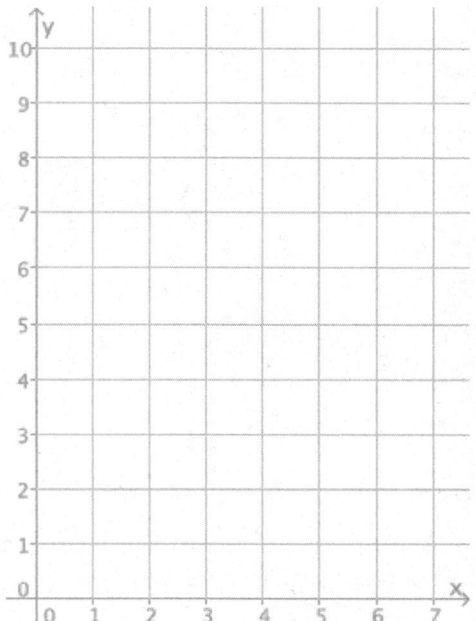

c. Plot the inputs and outputs as ordered pairs defining points on the coordinate plane.

d. What shape does the graph of the points appear to take?

e. Find the rate of change using rows 1 and 2 from the table above.

f. Find the rate of change using rows 2 and 3 from the table above.

g. Find the rate of change using any two other rows from the table above.

h. Return to your initial claim about the function. Is it linear or nonlinear? Justify your answer with as many pieces of evidence as possible.

## Exercises 4–10

In each of Exercises 4–10, an equation describing a rule for a function is given, and a question is asked about it. If necessary, use a table to organize pairs of inputs and outputs, and then plot each on a coordinate plane to help answer the question.

4. What shape do you expect the graph of the function described by $y = x$ to take? Is it a linear or nonlinear function?

5. What shape do you expect the graph of the function described by $y = 2x^2 - x$ to take? Is it a linear or nonlinear function?

6. What shape do you expect the graph of the function described by $3x + 7y = 8$ to take? Is it a linear or nonlinear function?

7. What shape do you expect the graph of the function described by $y = 4x^3$ to take? Is it a linear or nonlinear function?

8. What shape do you expect the graph of the function described by $\frac{3}{x} = y$ to take? Is it a linear or nonlinear function? (Assume that an input of $x = 0$ is disallowed.)

9. What shape do you expect the graph of the function described by $\frac{4}{x^2} = y$ to take? Is it a linear or nonlinear function? (Assume that an input of $x = 0$ is disallowed.)

10. What shape do you expect the graph of the equation $x^2 + y^2 = 36$ to take? Is it a linear or nonlinear function? Is it a function? Explain.

### Lesson Summary

One way to determine if a function is linear or nonlinear is to inspect average rates of change using a table of values. If these average rates of change are not constant, then the function is not linear.

Another way is to examine the graph of the function. If all the points on the graph do not lie on a common line, then the function is not linear.

If a function is described by an equation different from one equivalent to $y = mx + b$ for some fixed values $m$ and $b$, then the function is not linear.

Name ____________________________________________ Date ____________________

1. The graph below is the graph of a function. Do you think the function is linear or nonlinear? Briefly justify your answer.

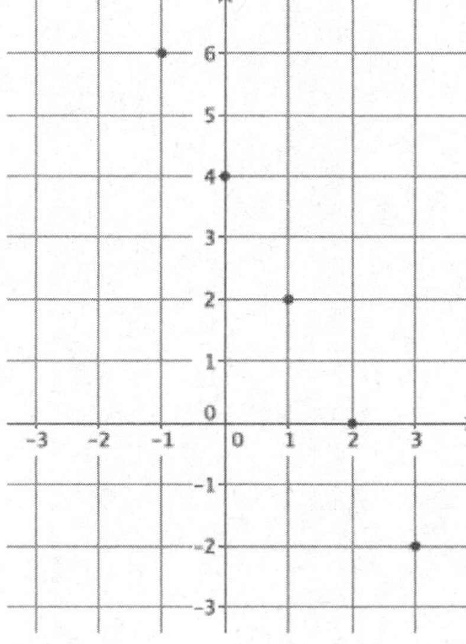


2. Consider the function that assigns to each number $x$ the value $\frac{1}{2}x^2$. Do you expect the graph of this function to be a straight line? Briefly justify your answer.

1. A function has the rule so that each input of $x$ is assigned an output of $x^3 + 1$.

> In Module 4, we analyzed linear and nonlinear equations. I remember that to be linear, the exponent of the variable, $x$, had to be equal to $1$.

a. Do you think the function is linear or nonlinear? Explain.

***The equation that describes this function is nonlinear because the exponent of the variable, $x$, is not equal to $1$, so I think this is a nonlinear function.***

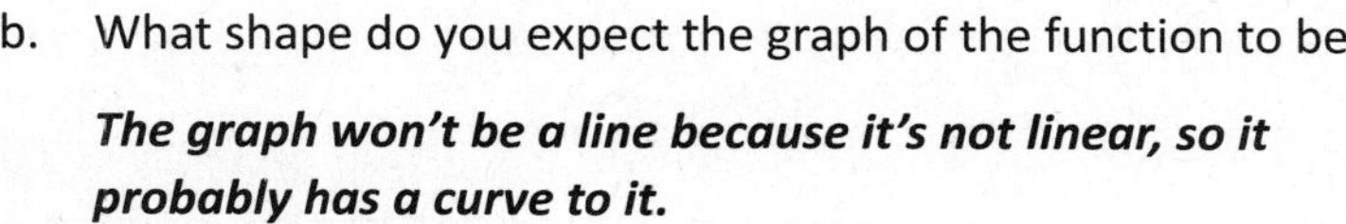

b. What shape do you expect the graph of the function to be?

***The graph won't be a line because it's not linear, so it probably has a curve to it.***

c. Develop a list of inputs and outputs for this function. Plot the inputs and outputs as points on the coordinate plane where the output is the $y$-coordinate.

| Input ($x$) | Output ($x^3 + 1$) |
|---|---|
| $-2$ | $(-2)^3 + 1 = -8 + 1 = -7$ |
| $-1$ | $(-1)^3 + 1 = -1 + 1 = 0$ |
| $0$ | $0^3 + 1 = 0 + 1 = 1$ |
| $1$ | $1^3 + 1 = 1 + 1 = 2$ |
| $2$ | $2^3 + 1 = 8 + 1 = 9$ |

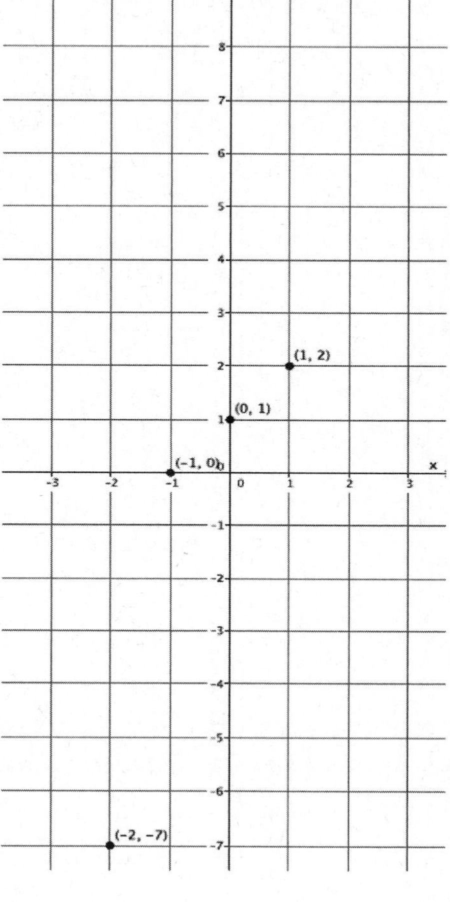


d. Was your prediction correct?

***I was right! There is no way to draw one straight line through all of the points on the graph. Therefore, this function is nonlinear.***

2. Is the function that is represented by this graph linear or nonlinear? Explain. Show work that supports your claim.

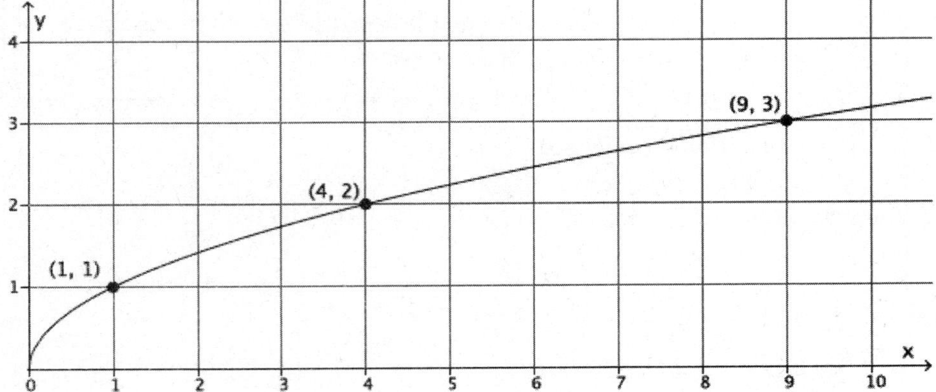


This graph doesn't look linear at all. The way to show work to support my claim is by showing that the rate of change between each pair of points is not equal to the same value.

$$\frac{3-2}{9-4} = \frac{1}{5}$$

$$\frac{2-1}{4-1} = \frac{1}{3}$$

$$\frac{3-1}{9-1} = \frac{2}{8} = \frac{1}{4}$$

***Since the rate of change was equal to a different value for all three pairs of inputs and outputs that were checked, the function represented by this graph is a nonlinear function.***

1. Consider the function that assigns to each number $x$ the value $x^2 - 4$.
   a. Do you think the function is linear or nonlinear? Explain.
   b. Do you expect the graph of this function to be a straight line?
   c. Develop a list of inputs and matching outputs for this function. Use them to begin a graph of the function.
   d. Was your prediction to (b) correct?

| **Input** $(x)$ | **Output** $(x^2 - 4)$ |
|---|---|
| $-3$ | |
| $-2$ | |
| $-1$ | |
| $0$ | |
| $1$ | |
| $2$ | |
| $3$ | |

2. Consider the function that assigns to each number $x$ greater than $-3$ the value $\frac{1}{x+3}$.
   a. Is the function linear or nonlinear? Explain.
   b. Do you expect the graph of this function to be a straight line?
   c. Develop a list of inputs and matching outputs for this function. Use them to begin a graph of the function.
   d. Was your prediction to (b) correct?

| **Input** $(x)$ | **Output** $\left(\frac{1}{x+3}\right)$ |
|---|---|
| $-2$ | |
| $-1$ | |
| $0$ | |
| $1$ | |
| $2$ | |
| $3$ | |

3.

a. Is the function represented by this graph linear or nonlinear? Briefly justify your answer.

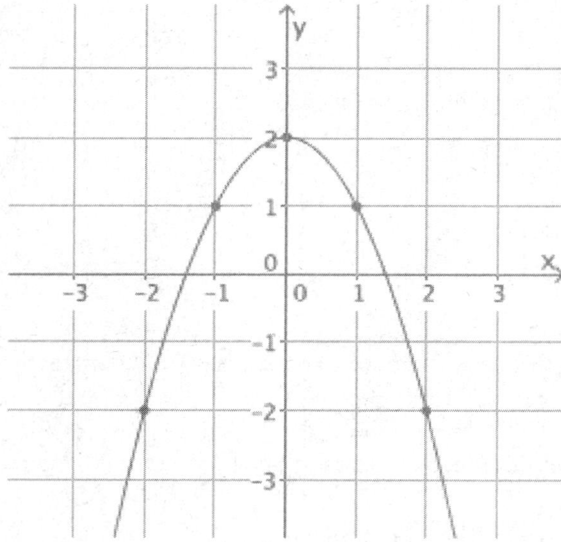


b. What is the average rate of change for this function from an input of $x = -2$ to an input of $x = -1$?

c. What is the average rate of change for this function from an input of $x = -1$ to an input of $x = 0$?

**Exploratory Challenge 1/Exercises 1–4**

As you complete Exercises 1–4, record the information in the table below.

<table>
<tr><th></th><th>Side length in inches ($s$)</th><th>Area in square inches ($A$)</th><th>Expression that describes area of border</th></tr>
<tr><td rowspan="2">Exercise 1</td><td></td><td></td><td rowspan="2"></td></tr>
<tr><td></td><td></td></tr>
<tr><td rowspan="2">Exercise 2</td><td></td><td></td><td rowspan="2"></td></tr>
<tr><td></td><td></td></tr>
<tr><td rowspan="2">Exercise 3</td><td></td><td></td><td rowspan="2"></td></tr>
<tr><td></td><td></td></tr>
<tr><td rowspan="2">Exercise 4</td><td></td><td></td><td rowspan="2"></td></tr>
<tr><td></td><td></td></tr>
</table>

1. Use the figure below to answer parts (a)–(f).

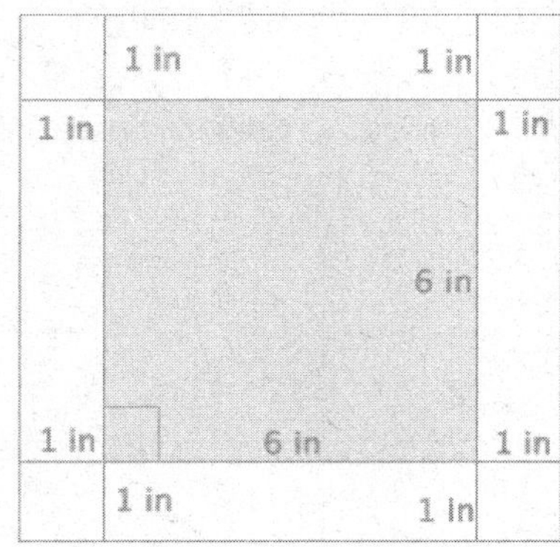


a. What is the length of one side of the smaller, inner square?

b. What is the area of the smaller, inner square?

c. What is the length of one side of the larger, outer square?

d. What is the area of the larger, outer square?

e. Use your answers in parts (b) and (d) to determine the area of the 1-inch white border of the figure.

f. Explain your strategy for finding the area of the white border.

2. Use the figure below to answer parts (a)–(f).

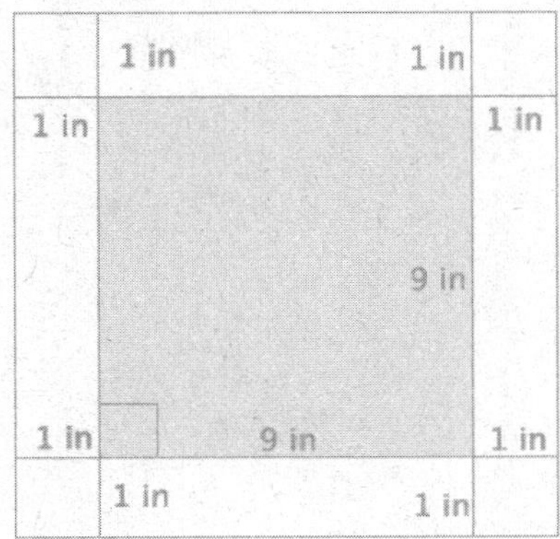


a. What is the length of one side of the smaller, inner square?

b. What is the area of the smaller, inner square?

c. What is the length of one side of the larger, outer square?

d. What is the area of the larger, outer square?

e. Use your answers in parts (b) and (d) to determine the area of the 1-inch white border of the figure.

f. Explain your strategy for finding the area of the white border.

3. Use the figure below to answer parts (a)–(f).

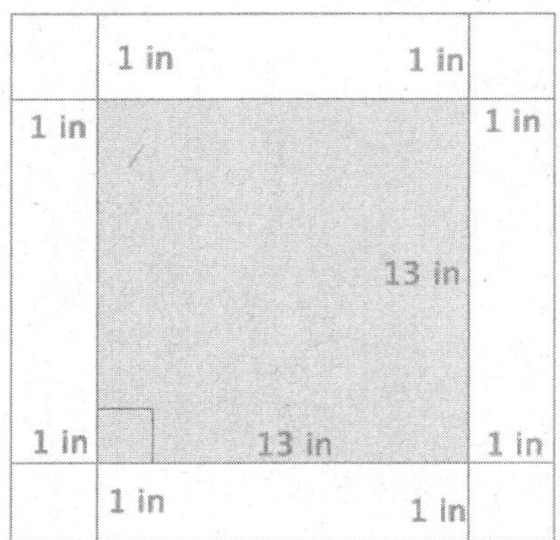


a. What is the length of one side of the smaller, inner square?

b. What is the area of the smaller, inner square?

c. What is the length of one side of the larger, outer square?

d. What is the area of the larger, outer square?

e. Use your answers in parts (b) and (d) to determine the area of the 1-inch white border of the figure.

f. Explain your strategy for finding the area of the white border.

4. Write a function that would allow you to calculate the area of a 1-inch white border for any sized square picture measured in inches.

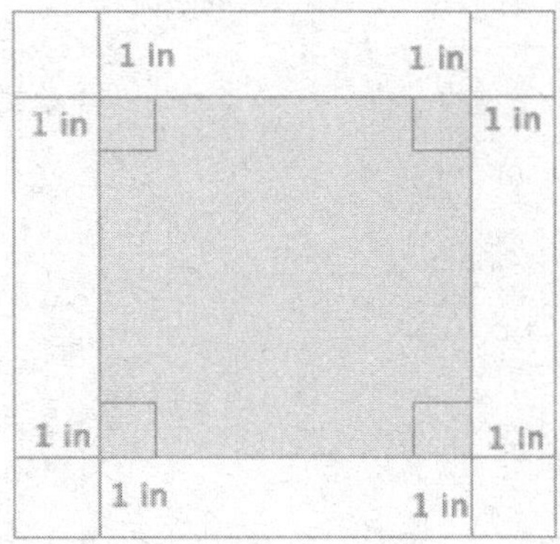


a. Write an expression that represents the side length of the smaller, inner square.

b. Write an expression that represents the area of the smaller, inner square.

c. Write an expression that represents the side lengths of the larger, outer square.

d. Write an expression that represents the area of the larger, outer square.

e. Use your expressions in parts (b) and (d) to write a function for the area $A$ of the 1-inch white border for any sized square picture measured in inches.

**Exercises 5–6**

5. The volume of the prism shown below is $61.6\text{ in}^3$. What is the height of the prism?

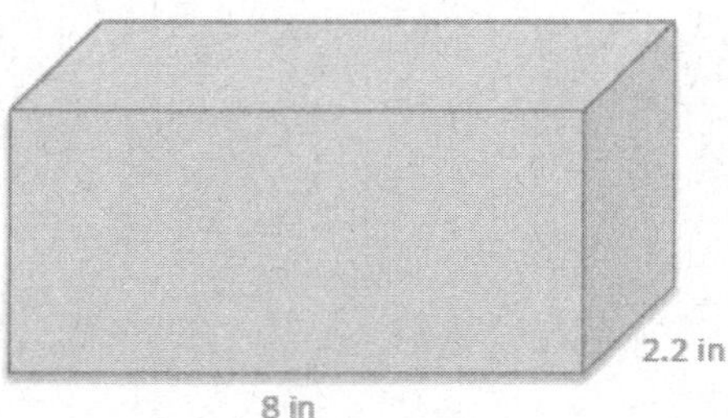


6. Find the value of the ratio that compares the volume of the larger prism to the smaller prism.

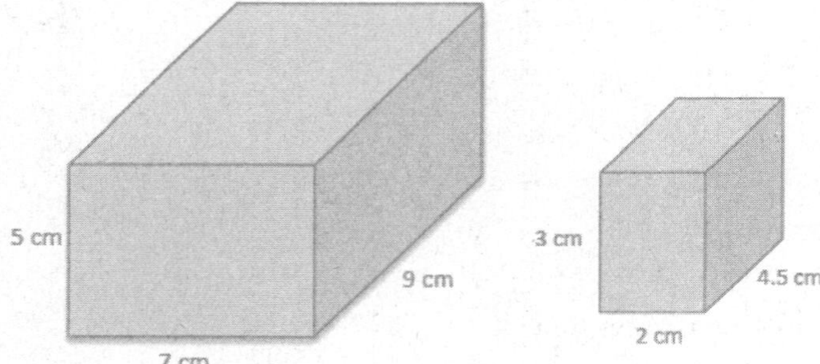

**Exploratory Challenge 2/Exercises 7–10**

As you complete Exercises 7–10, record the information in the table below. Note that *base* refers to the bottom of the prism.

| | Area of base in square centimeters ($B$) | Height in centimeters ($h$) | Volume in cubic centimeters |
|---|---|---|---|
| **Exercise 7** | | | |
| **Exercise 8** | | | |
| **Exercise 9** | | | |
| **Exercise 10** | | | |

7. Use the figure to the right to answer parts (a)–(c).

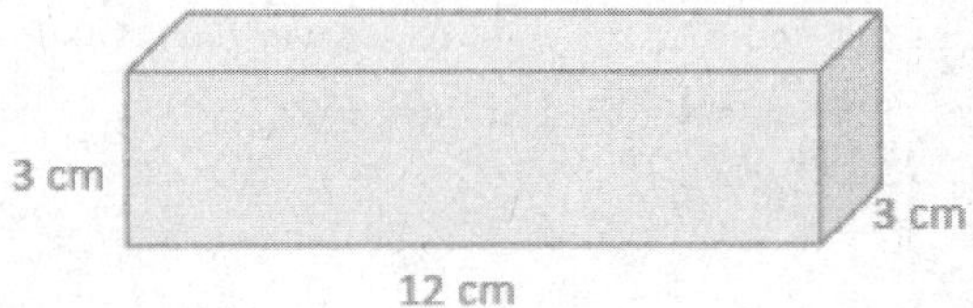


   a. What is the area of the base?

   b. What is the height of the figure?

   c. What is the volume of the figure?

8. Use the figure to the right to answer parts (a)–(c).

   a. What is the area of the base?

   b. What is the height of the figure?

   c. What is the volume of the figure?

9. Use the figure to the right to answer parts (a)–(c).

   a. What is the area of the base?

   b. What is the height of the figure?

   c. What is the volume of the figure?

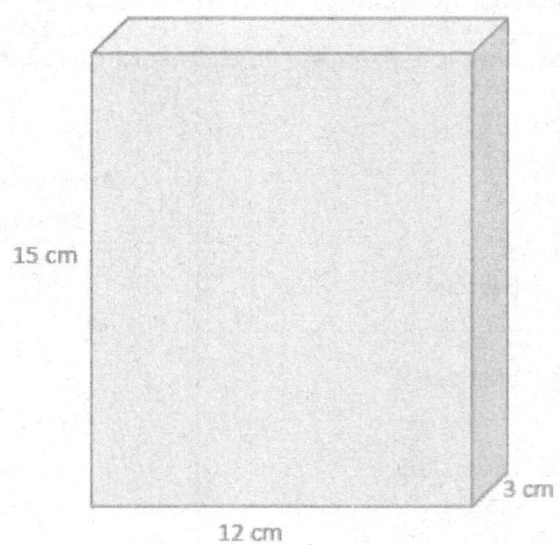


10. Use the figure to the right to answer parts (a)–(c).

   a. What is the area of the base?

   b. What is the height of the figure?

   c. Write and describe a function that will allow you to determine the volume of any rectangular prism that has a base area of $36 \text{ cm}^2$.

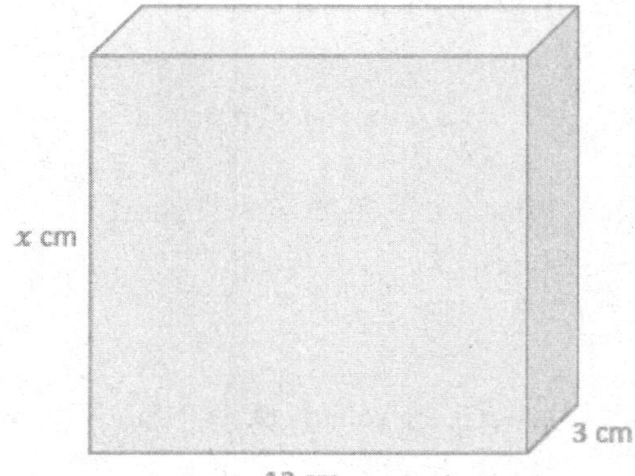

## Lesson Summary

There are a few basic assumptions that are made when working with volume:

(a) The volume of a solid is always a number greater than or equal to $0$.

(b) The volume of a unit cube (i.e., a rectangular prism whose edges all have a length of $1$) is by definition $1$ cubic unit.

(c) If two solids are identical, then their volumes are equal.

(d) If two solids have (at most) their boundaries in common, then their total volume can be calculated by adding the individual volumes together. (These figures are sometimes referred to as composite solids.)

Name ____________________________________ Date ____________________

1. Write a function that would allow you to calculate the area in square inches, $A$, of a 2-inch white border for any sized square figure with sides of length $s$ measured in inches.

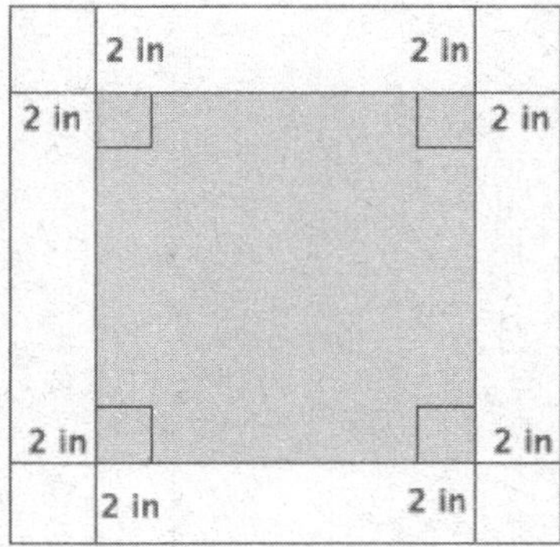


2. The volume of the rectangular prism is $295.68\ \text{in}^3$. What is its width?

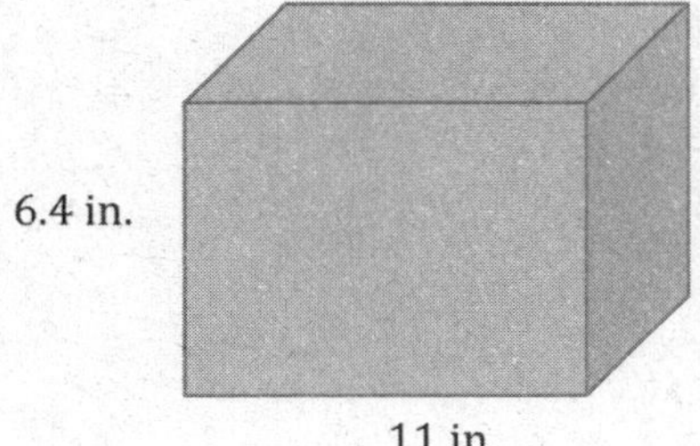

1. Write a function that would allow you to calculate the area, $A$, of a 5-inch thick outer ring for any sized dartboard with radius $r$ inches. Write an exact answer that uses $\pi$ (*do not* approximate your answer by using $3.14$ for $\pi$).

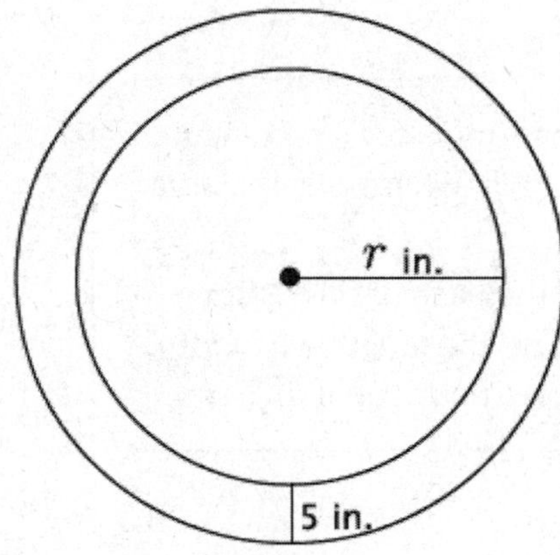


This is a lot like the problem we did in class. I need to write an expression that shows the difference of the areas in the two circles, the circle with radius $r$ in. and the circle with radius $(r + 5)$ in.

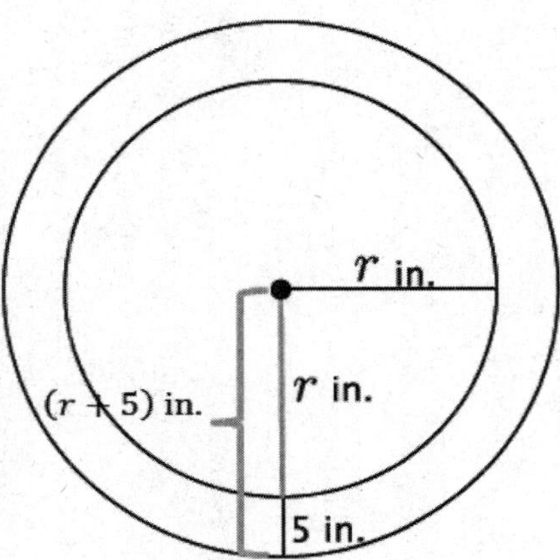


***The area of the inner, smaller, circle in square inches is found by calculating*** $\pi r^2$.

***The area of the outer, larger, circle in square inches is found by calculating*** $\pi(r + 5)^2$.

***To find the area of the outer ring, we have to find the difference of the two areas.***

$$A = \pi(r + 5)^2 - \pi r^2$$

***The area of the outer ring is*** $(\pi(r + 5)^2 - \pi r^2)\ \text{in}^2$.

2. The shell of the solid was filled with water and then poured into the standard rectangular prism, as shown. The height that the volume reaches is $74.68$ cm. What is the volume of the solid?

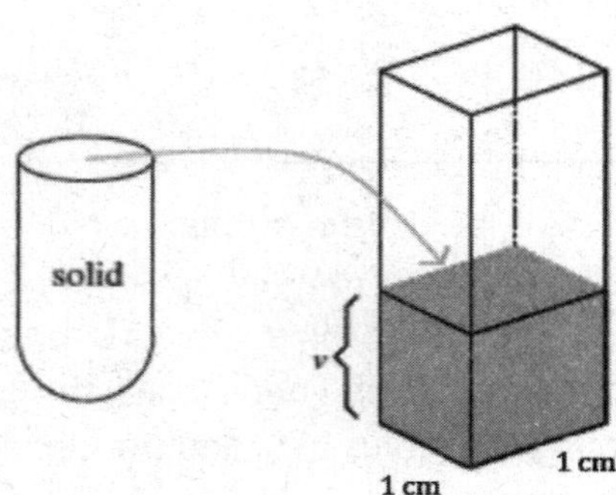


The volume formula for a rectangular prism is $V = Bh$, where $B$ is the area of the base.

The area of the base of this prism is $1\ \text{cm}^2$ because the length and width of the base are both equal to $1$ cm.

***B* = 1 *and h* = 74.68, *so***

$$V = (1)(74.68)$$
$$V = 74.68$$

***The volume is* $74.68\ \text{cm}^3$.**

3. The volume of the prism shown below is $60\ \text{in}^3$. What is its length?

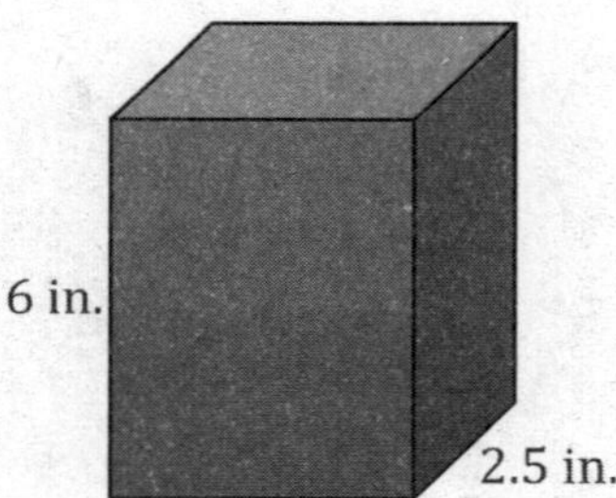


***The area of the base, B, is found by multiplying the length of the base, l, by the width of the base, w.***

$$V = Bh$$
$$V = l \times w \times h$$
$$60 = l \times 2.5 \times 6$$
$$60 = l \times 15$$
$$\frac{60}{15} = l\left(\frac{15}{15}\right)$$
$$4 = l$$

***The length of the prism is* $4$ in.**

1. Calculate the area of the 3-inch white border of the square figure below.

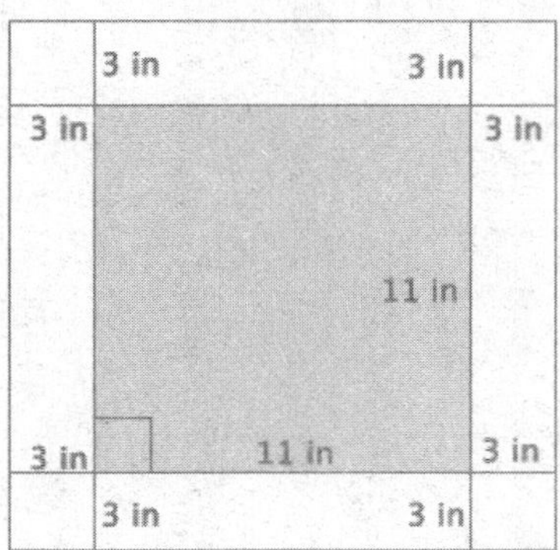


2. Write a function that would allow you to calculate the area, $A$, of a 3-inch white border for any sized square picture measured in inches.

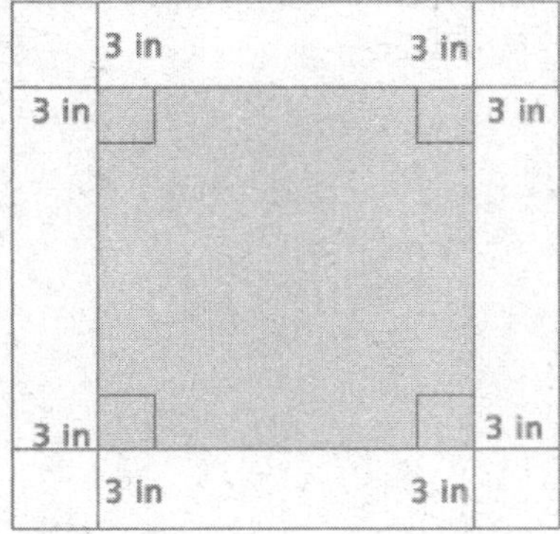


3. Dartboards typically have an outer ring of numbers that represent the number of points a player can score for getting a dart in that section. A simplified dartboard is shown below. The center of the circle is point $A$. Calculate the area of the outer ring. Write an exact answer that uses $\pi$ (*do not* approximate your answer by using 3.14 for $\pi$).

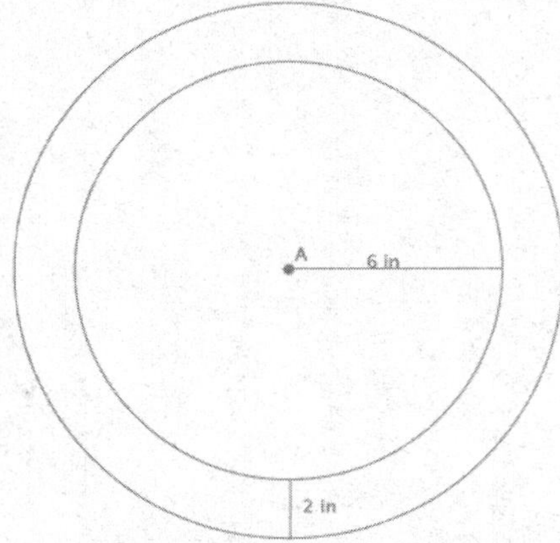

4. Write a function that would allow you to calculate the area, $A$, of the outer ring for any sized dartboard with radius $r$. Write an exact answer that uses $\pi$ (*do not* approximate your answer by using $3.14$ for $\pi$).

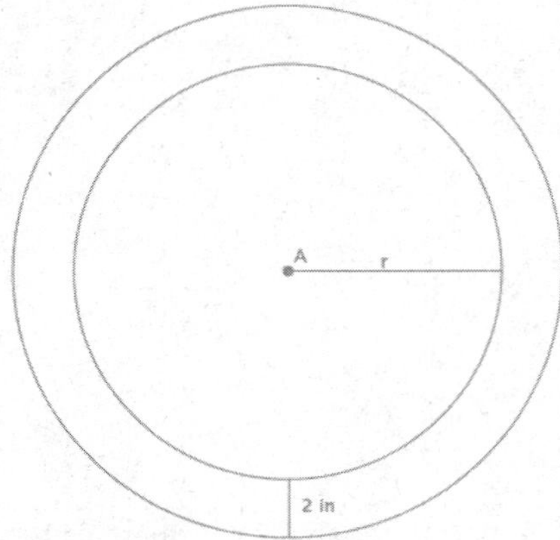


5. The shell of the solid shown was filled with water and then poured into the standard rectangular prism, as shown. The height that the volume reaches is $14.2$ in. What is the volume of the shell of the solid?

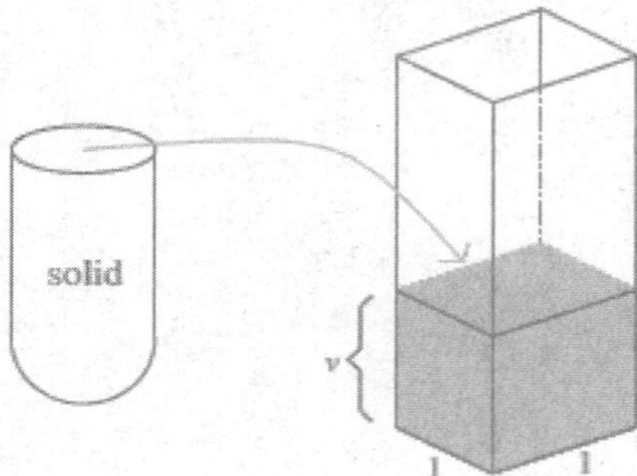


6. Determine the volume of the rectangular prism shown below.

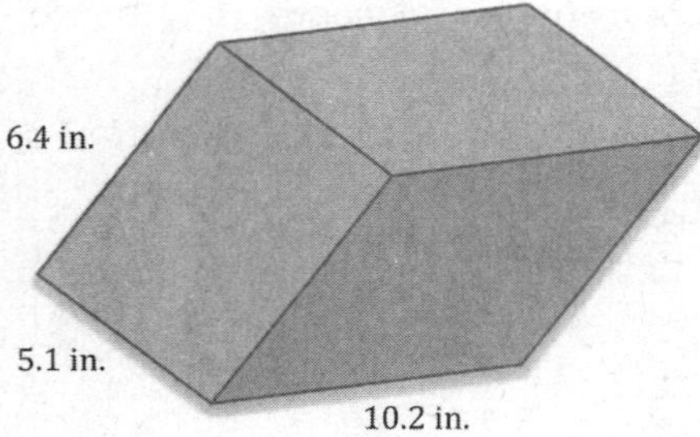

7. The volume of the prism shown below is $972\text{ cm}^3$. What is its length?

8. The volume of the prism shown below is $32.7375\text{ ft}^3$. What is its width?

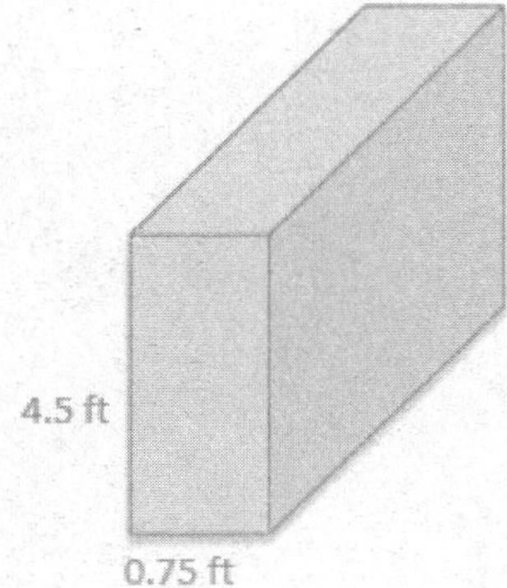


9. Determine the volume of the 3-dimensional figure below. Explain how you got your answer.

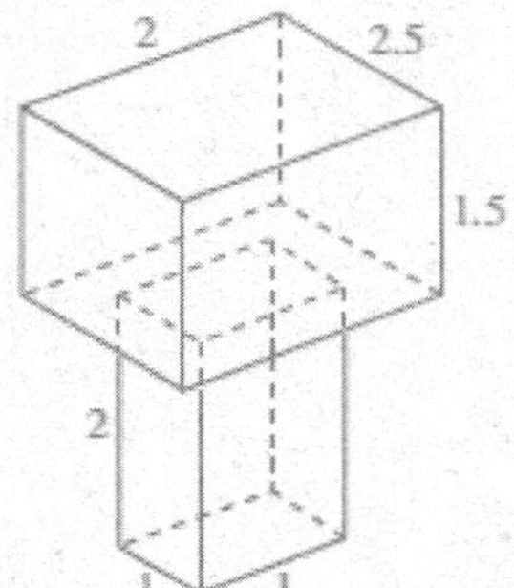

**Opening Exercise**

a.

i. Write an equation to determine the volume of the rectangular prism shown below.

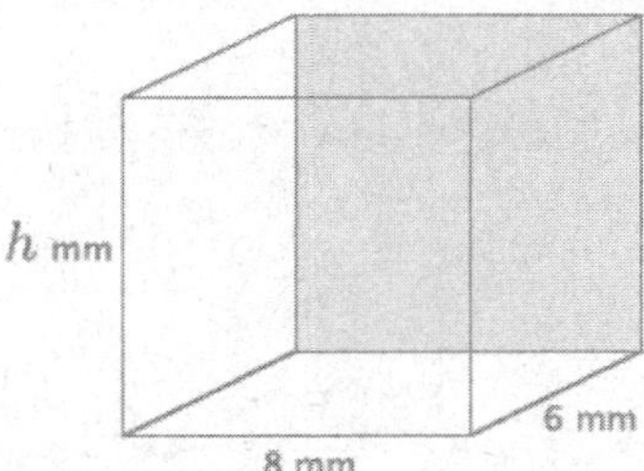


ii. Write an equation to determine the volume of the rectangular prism shown below.

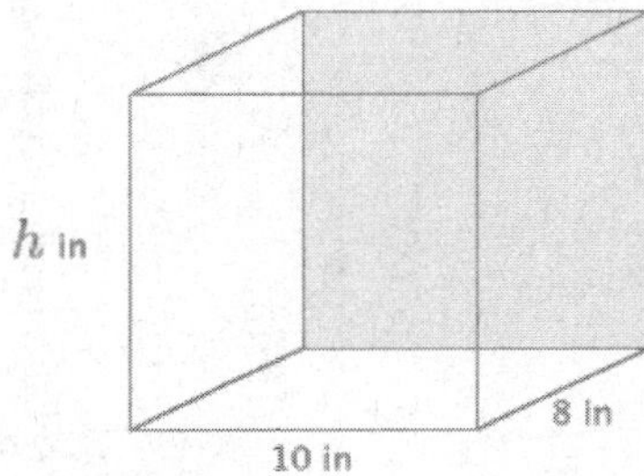


iii. Write an equation to determine the volume of the rectangular prism shown below.

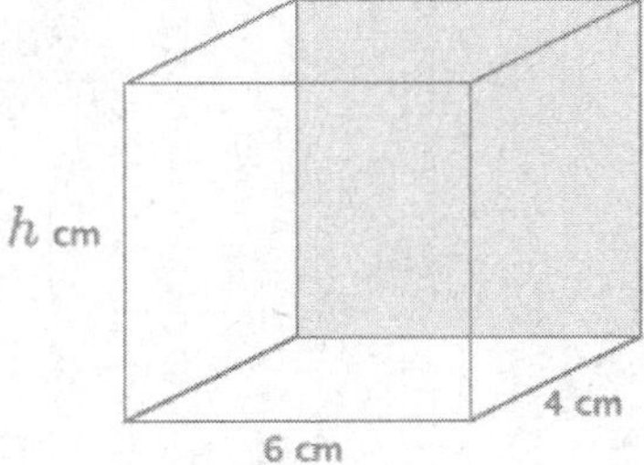

iv. Write an equation for volume, $V$, in terms of the area of the base, $B$.

b. Using what you learned in part (a), write an equation to determine the volume of the cylinder shown below.

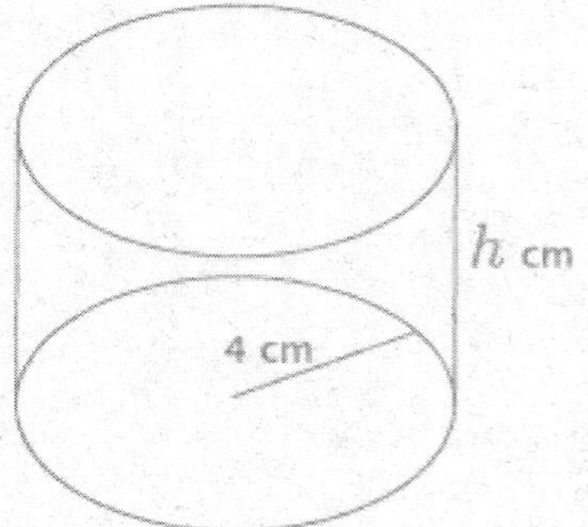


## Exercises 1–3

1. Use the diagram to the right to answer the questions.

   a. What is the area of the base?

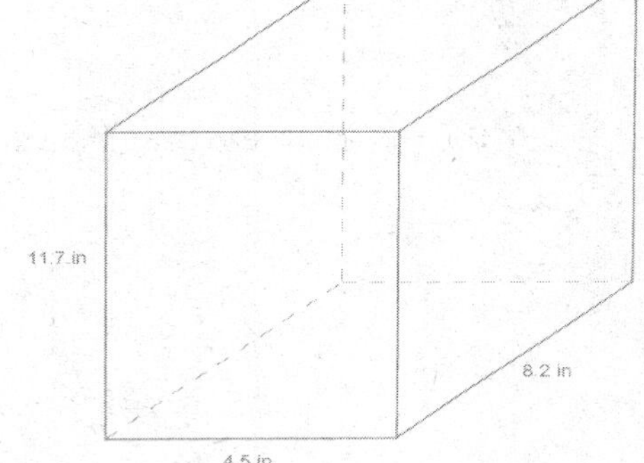


   b. What is the height?

   c. What is the volume of the rectangular prism?

2. Use the diagram to the right to answer the questions.

   a. What is the area of the base?

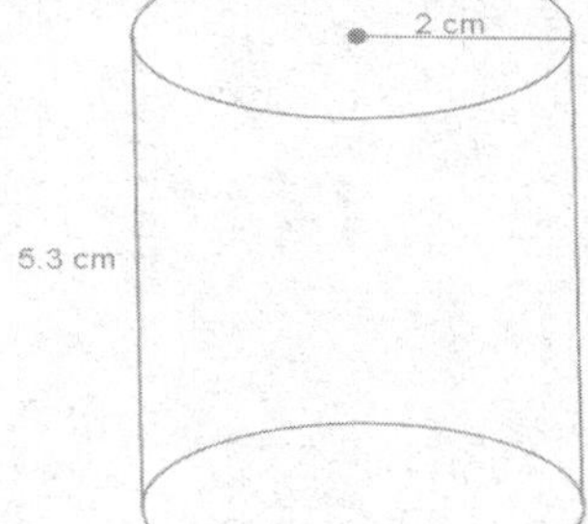


   b. What is the height?

   c. What is the volume of the right circular cylinder?

3. Use the diagram to the right to answer the questions.

   a. What is the area of the base?

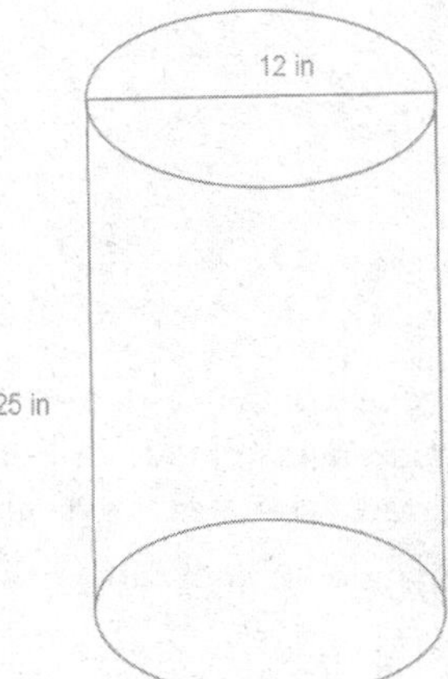


   b. What is the height?

   c. What is the volume of the right circular cylinder?

**Exercises 4–6**

4. Use the diagram to find the volume of the right circular cone.

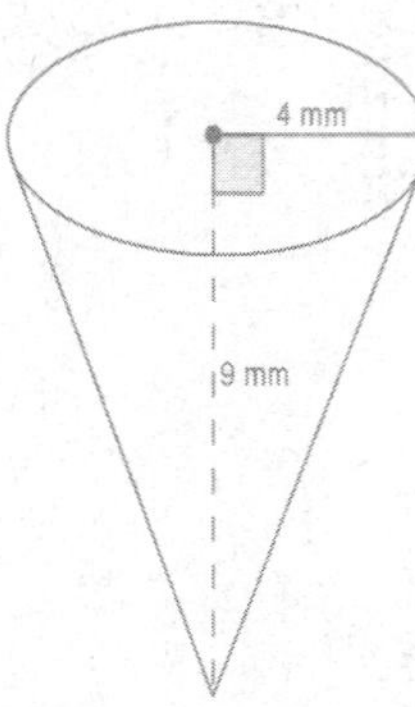


5. Use the diagram to find the volume of the right circular cone.

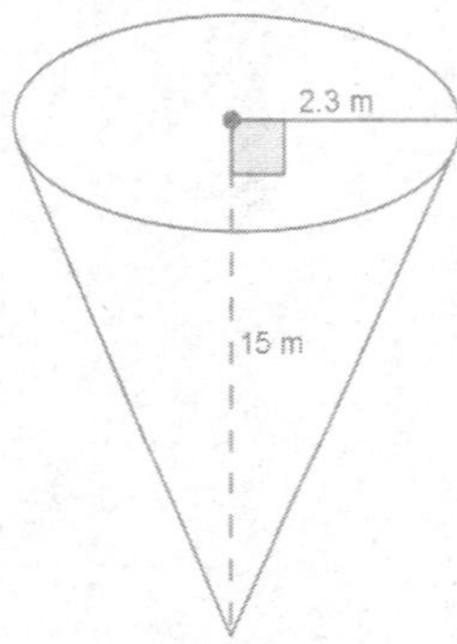


6. Challenge: A container in the shape of a right circular cone has height $h$, and base of radius $r$, as shown. It is filled with water (in its upright position) to half the height. Assume that the surface of the water is parallel to the base of the inverted cone. Use the diagram to answer the following questions:

   a. What do we know about the lengths of $AB$ and $AO$?

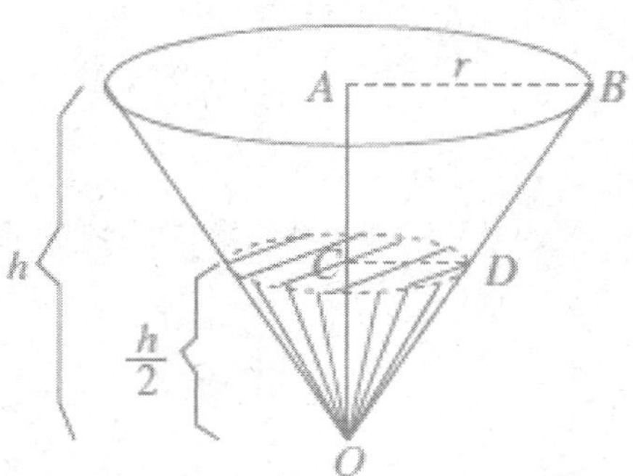

b. What do we know about the measure of $\angle OAB$ and $\angle OCD$?

c. What can you say about $\triangle\ OAB$ and $\triangle\ OCD$?

d. What is the ratio of the volume of water to the volume of the container itself?

## Lesson Summary

The formula to find the volume, $V$, of a right circular cylinder is $V = \pi r^2 h = Bh$, where $B$ is the area of the base.

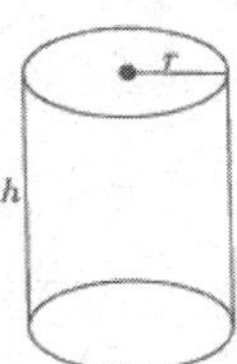


The formula to find the volume of a cone is directly related to that of the cylinder. Given a right circular cylinder with radius $r$ and height $h$, the volume of a cone with those same dimensions is one-third of the cylinder. The formula for the volume, $V$, of a circular cone is $V = \frac{1}{3}\pi r^2 h$. More generally, the volume formula for a general cone is $V = \frac{1}{3}Bh$, where $B$ is the area of the base.

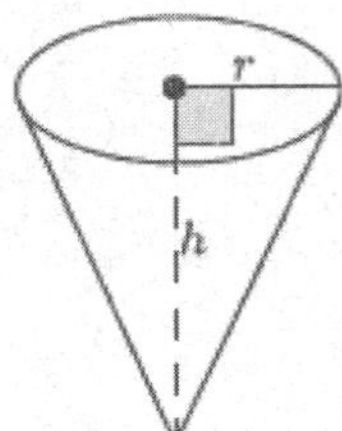

Name ______________________________ Date ______________

1. Use the diagram to find the total volume of the three cones shown below.

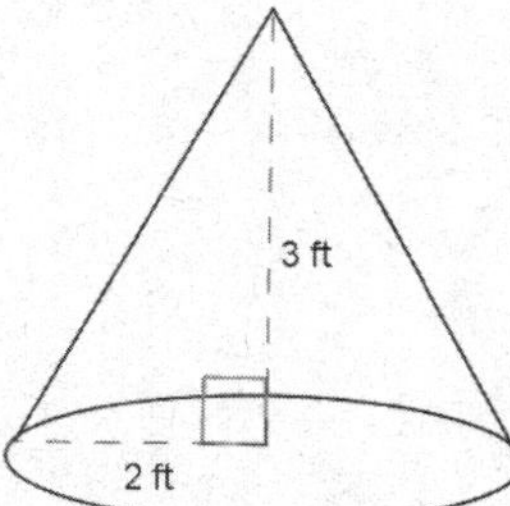

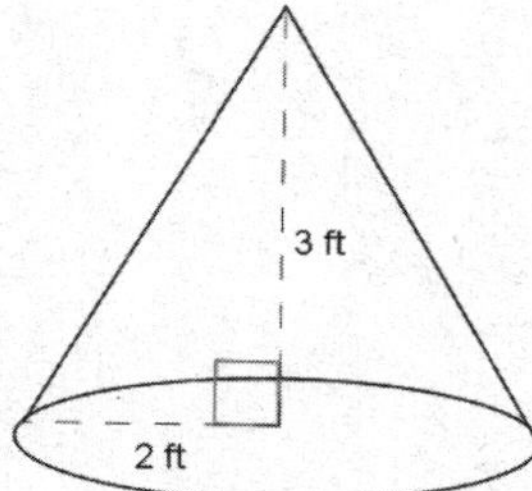

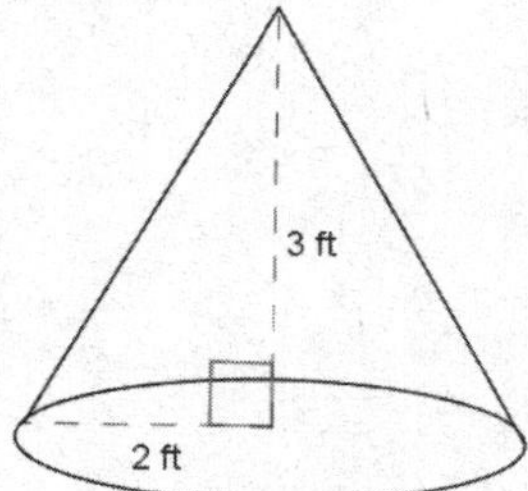


2. Use the diagram below to determine which has the greater volume, the cone or the cylinder.

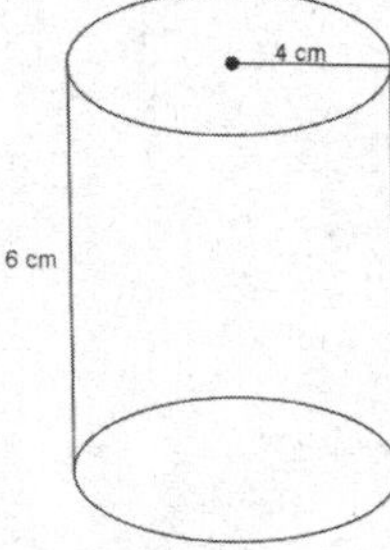

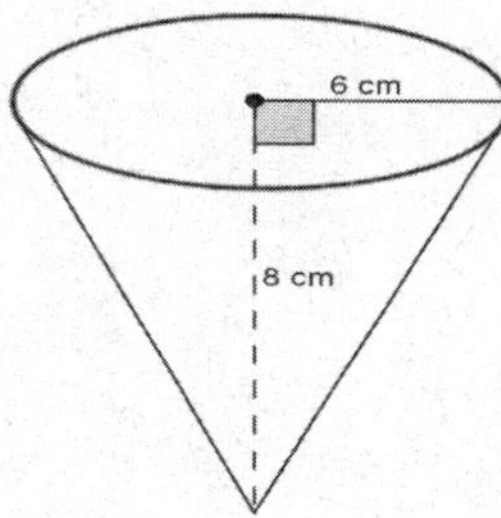

1. Dayna wants to add water to a bucket that is the shape of a right circular cylinder. The bucket has a 4-inch radius and 8-inch height. She uses a scoop that has the shape of right circular cone with a 2-inch radius and 3-inch height. How many scoops will it take Dayna to fill the bucket up level with the top?

$V = \pi r^2 h$

$V = \pi(4)^2(8)$

$V = 128\pi$

If I take the volume of the cylinder and divide it by the volume of the cone, the answer will be the number of scoops it takes Dayna to fill that cylinder.

***The volume of the cylinder is*** $128\pi \text{ in}^3$.

$V = \frac{1}{3}\pi r^2 h$

$V = \frac{1}{3}\pi(2)^2(3)$

$V = 4\pi$

***The volume of the cone is*** $4\pi \text{ in}^3$.

$\frac{128\pi}{4\pi} = 32$

***The number of scoops needed to fill the cylinder is*** $32$.

2. A cylindrical tank (with dimensions shown below) contains water that is 5 feet deep. If water is poured into the tank at a constant rate of 24 $\frac{\text{ft}^3}{\text{min}}$ for 18 min., will the tank overflow? Use 3.14 to estimate $\pi$.

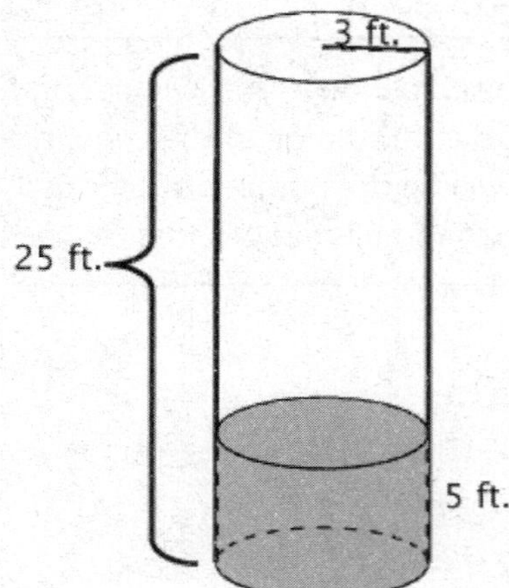


I need to figure out the total volume of the cylinder and what is already filled. Then, I need to compare the difference to the amount of water that is being poured into the tank.

$$V = \pi r^2 h$$
$$V = \pi(3)^2(25)$$
$$V \approx (3.14)(9)(25)$$
$$V \approx 706.5$$

***The volume of the tank is approximately*** $\mathbf{706.5\ ft^3}$**.**

$$V = \pi r^2 h$$
$$V = \pi(3)^2(5)$$
$$V \approx (3.14)(9)(5)$$
$$V \approx 141.3$$

***The volume of the part of the tank that is already filled is approximately*** $\mathbf{141.3\ ft^3}$**.**

$$706.5 - 141.3 = 565.2$$

***The volume of the tank that is not filled is approximately*** $\mathbf{565.2\ ft^3}$**.**

$$24 \times 18 = 432$$

***The volume of water poured into the tank is*** $\mathbf{432\ ft^3}$**.**

***Since the volume of water going into the tank is less than the volume that remains in the tank, it will not overflow.***

1. Use the diagram to help you find the volume of the right circular cylinder.

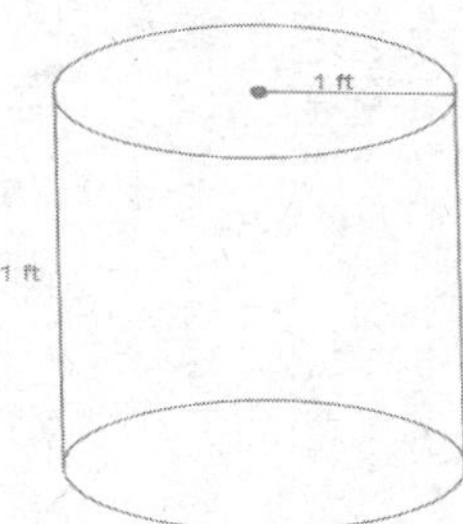


2. Use the diagram to help you find the volume of the right circular cone.

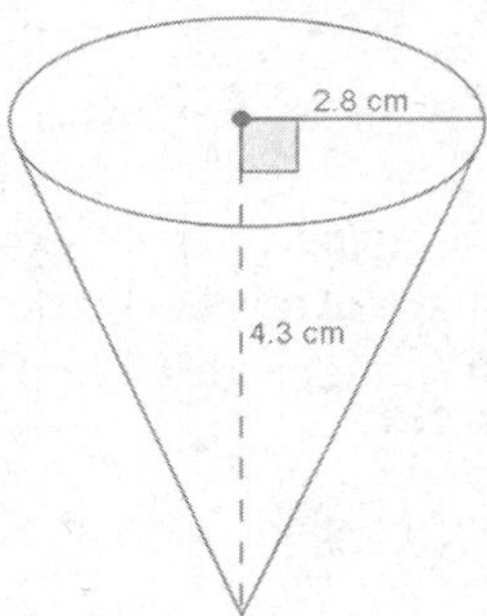


3. Use the diagram to help you find the volume of the right circular cylinder.

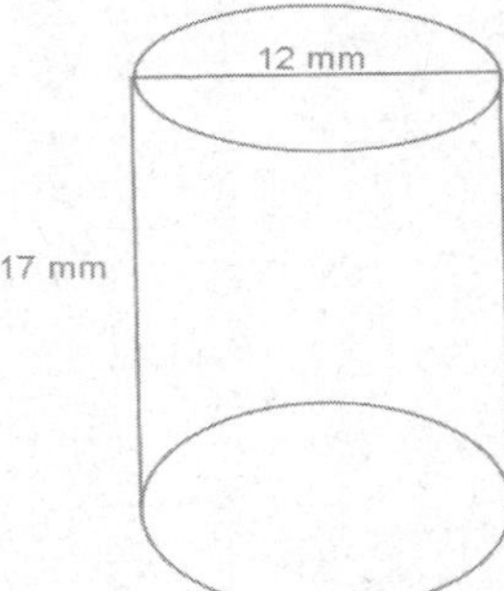

4. Use the diagram to help you find the volume of the right circular cone.

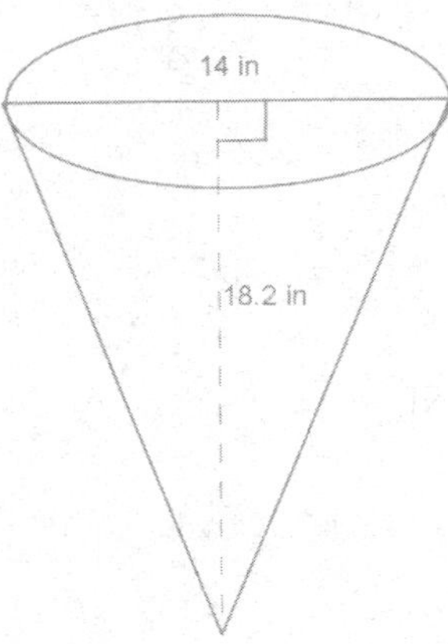


5. Oscar wants to fill with water a bucket that is the shape of a right circular cylinder. It has a 6-inch radius and 12-inch height. He uses a shovel that has the shape of a right circular cone with a 3-inch radius and 4-inch height. How many shovelfuls will it take Oscar to fill the bucket up level with the top?

6. A cylindrical tank (with dimensions shown below) contains water that is 1-foot deep. If water is poured into the tank at a constant rate of $20\ \frac{\text{ft}^3}{\text{min}}$ for $20$ min., will the tank overflow? Use $3.14$ to estimate $\pi$.

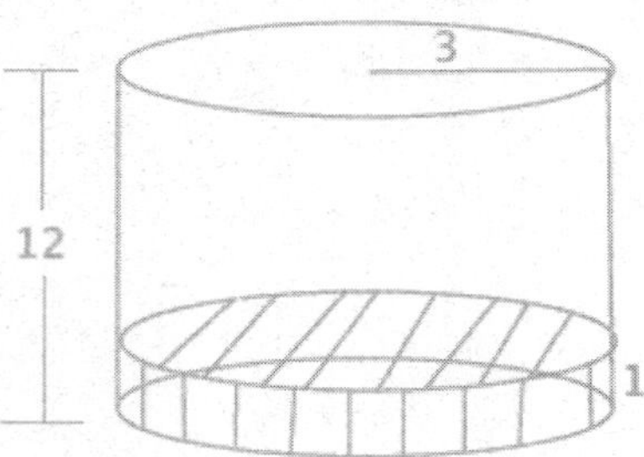

**Exercises 1–3**

1. What is the volume of a cylinder?

2. What is the height of the cylinder?

3. If $\text{volume(sphere)} = \frac{2}{3}\text{volume(cylinder with same diameter and height)}$, what is the formula for the volume of a sphere?

Example 1

Compute the exact volume for the sphere shown below.

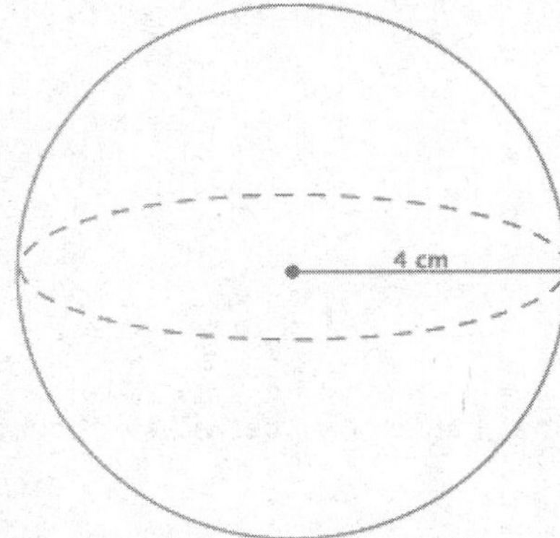

## Example 2

A cylinder has a diameter of 16 inches and a height of 14 inches. What is the volume of the largest sphere that will fit into the cylinder?

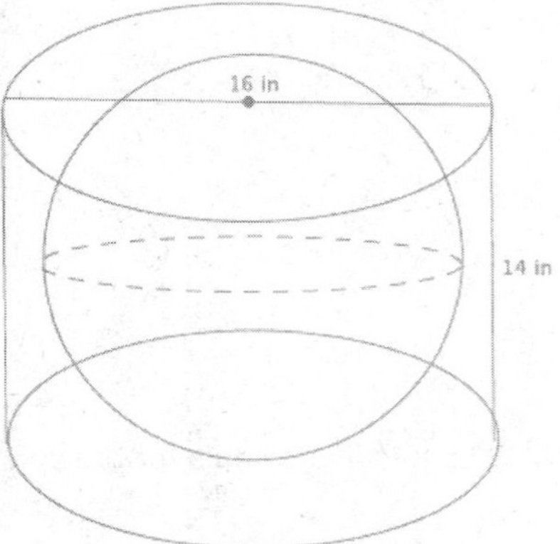


### Exercises 4–8

4. Use the diagram and the general formula to find the volume of the sphere.

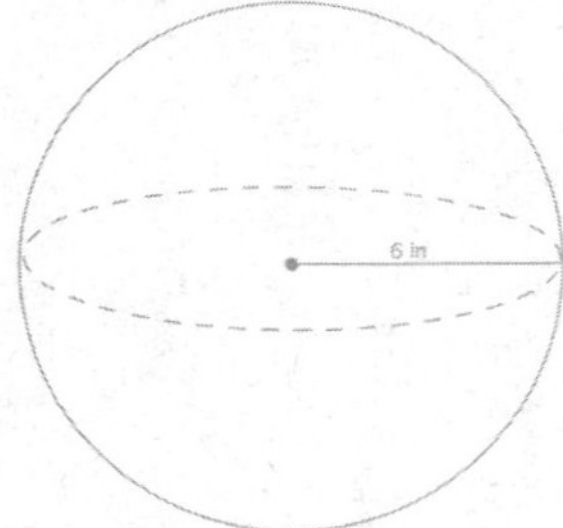


5. The average basketball has a diameter of 9.5 inches. What is the volume of an average basketball? Round your answer to the tenths place.

6. A spherical fish tank has a radius of 8 inches. Assuming the entire tank could be filled with water, what would the volume of the tank be? Round your answer to the tenths place.

7. Use the diagram to answer the questions.

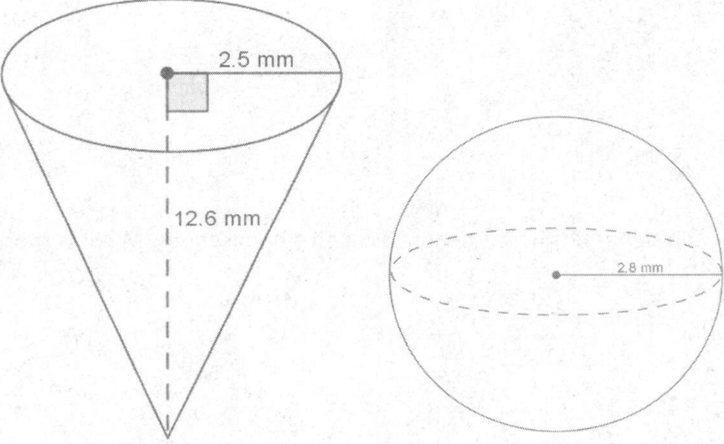


a. Predict which of the figures shown above has the greater volume. Explain.

b. Use the diagram to find the volume of each, and determine which has the greater volume.

8. One of two half spheres formed by a plane through the sphere's center is called a hemisphere. What is the formula for the volume of a hemisphere?

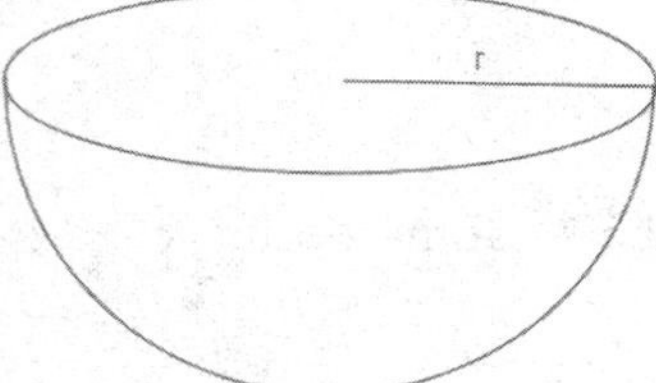

## Lesson Summary

The formula to find the volume of a sphere is directly related to that of the right circular cylinder. Given a right circular cylinder with radius $r$ and height $h$, which is equal to $2r$, a sphere with the same radius $r$ has a volume that is exactly two-thirds of the cylinder.

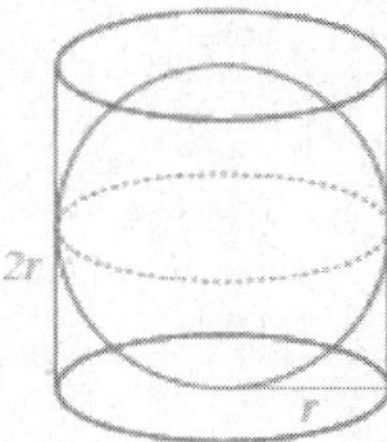


Therefore, the volume of a sphere with radius $r$ has a volume given by the formula $V = \frac{4}{3}\pi r^3$.

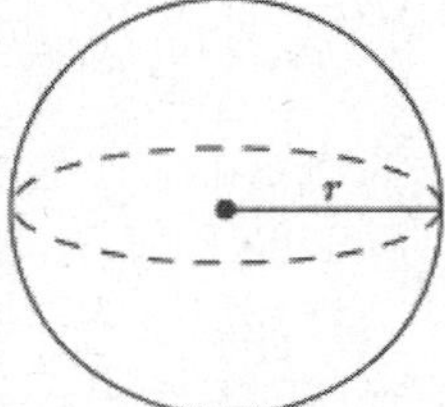

Name ____________________________________________ Date ____________________

1. What is the volume of the sphere shown below?

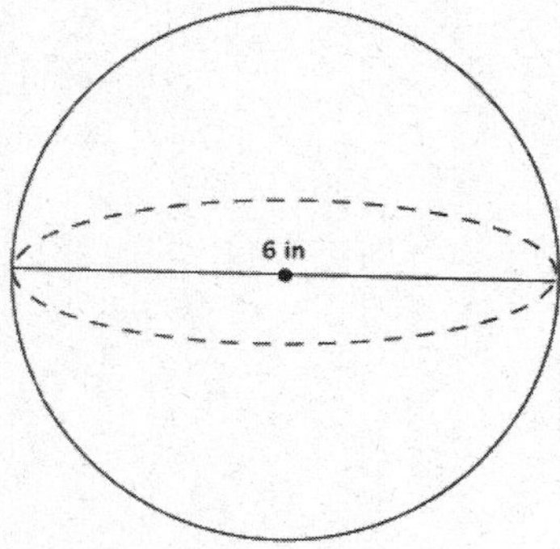


2. Which of the two figures below has the greater volume?

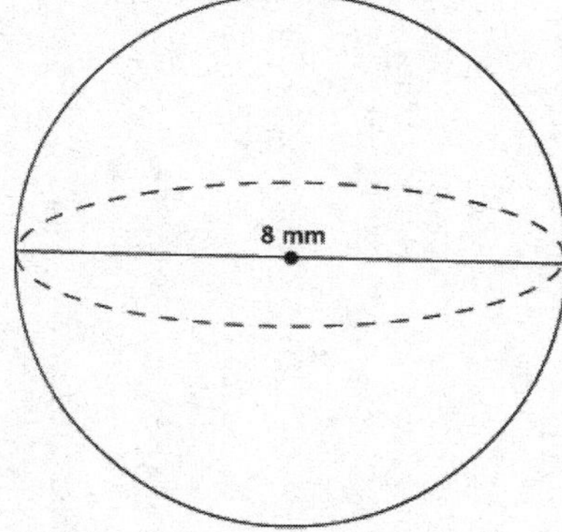


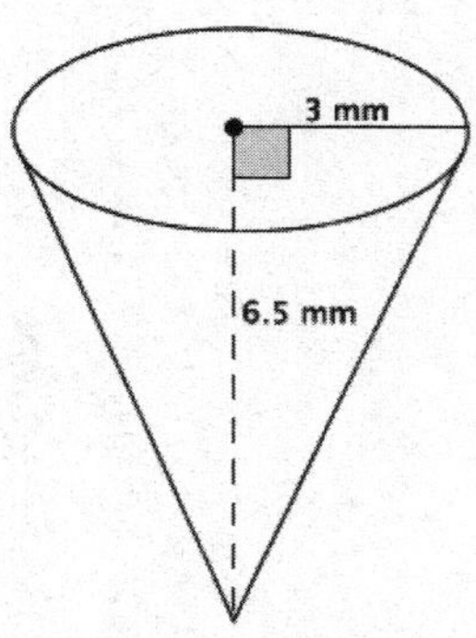

1. Which of the two figures below has the <u>lesser</u> volume? (Note: Figures are not drawn to scale.)

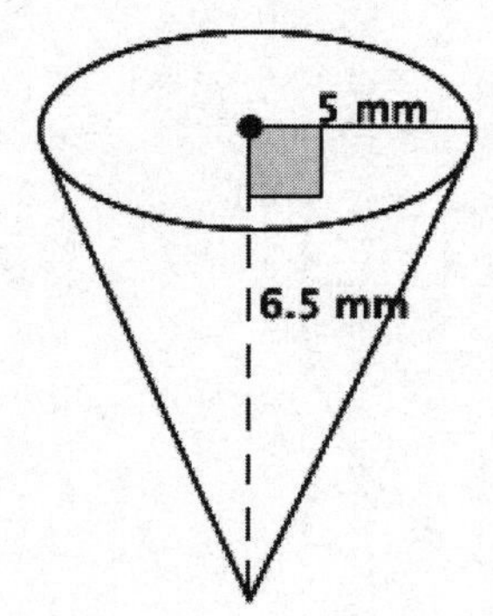


The volume formula for the sphere is $V = \frac{4}{3}\pi r^3$. All I have to do is calculate the volume of each figure and then compare.

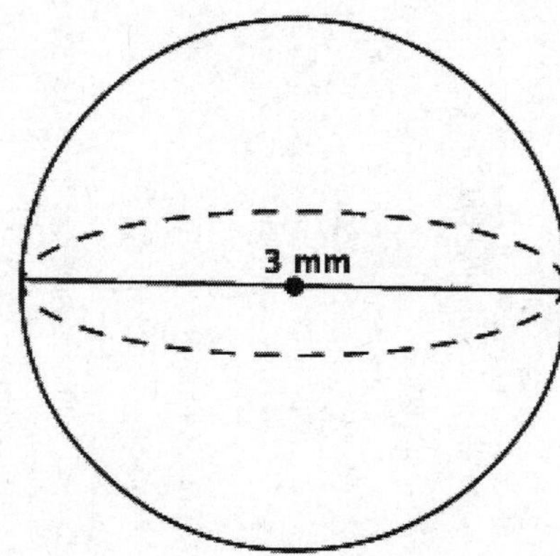


$$V = \frac{1}{3}\pi(5)^2(6.5)$$

$$V = \frac{162.5}{3}\pi$$

***The volume of the cone is*** $\frac{162.5}{3}\pi\ \text{mm}^3$.

$$V = \frac{4}{3}\pi(1.5)^3$$

$$V = \frac{13.5}{3}\pi$$

***The volume of the sphere is*** $\frac{13.5}{3}\pi\ \text{mm}^3$.

***Since*** $\frac{13.5}{3}\pi < \frac{162.5}{3}\pi$, ***the sphere has the lesser volume.***

2. Which of the two figures below has the greater volume? (Note: Figures are not drawn to scale.)

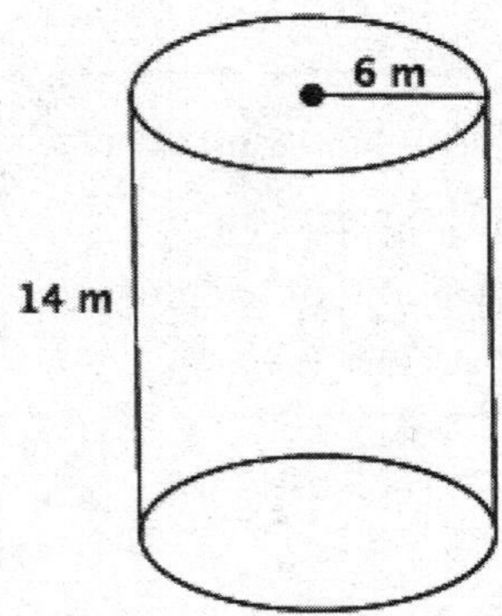


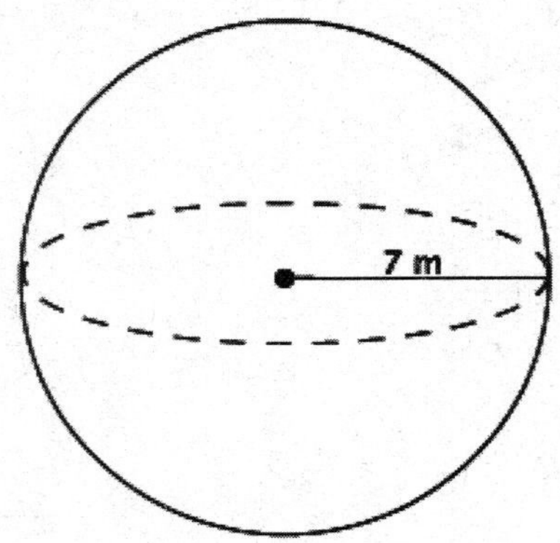


$$V = \pi(6)^2(14)$$
$$V = 504\pi$$

***The volume of the cylinder is*** $504\pi\ \text{m}^3$.

$$V = \frac{4}{3}\pi(7)^3$$
$$V = \frac{1372}{3}\pi$$
$$V \approx 457.3\pi$$

***The volume of the sphere is approximately*** $457.3\pi\ \text{m}^3$.

***Since*** $504\pi > 457.3\pi$, ***the cylinder has the greater volume.***

1. Use the diagram to find the volume of the sphere.

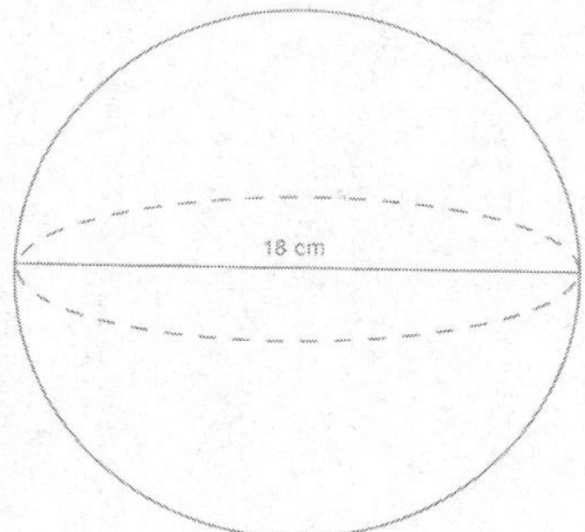


2. Determine the volume of a sphere with diameter 9 mm, shown below.

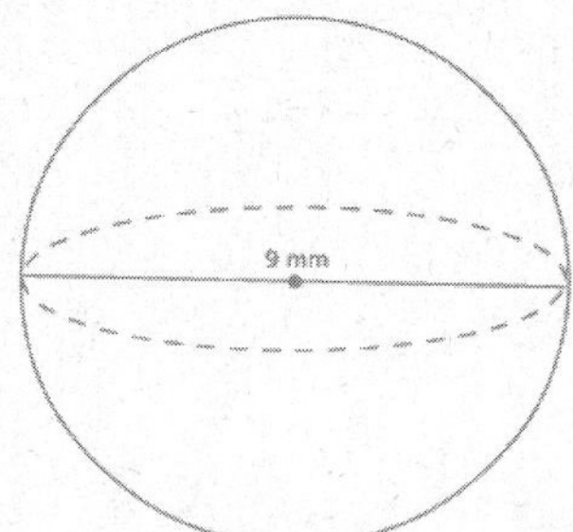


3. Determine the volume of a sphere with diameter 22 in., shown below.

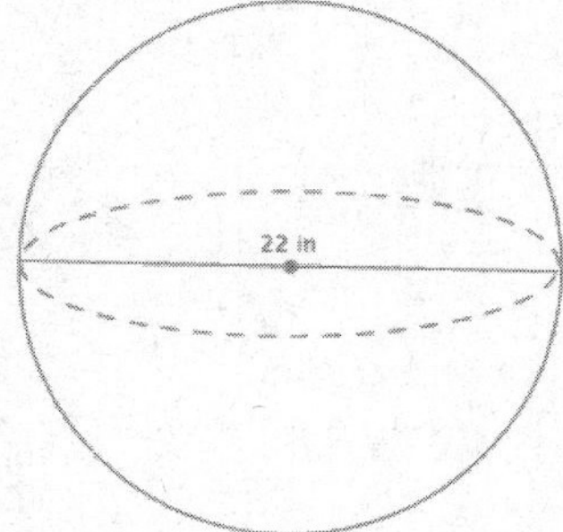

4. Which of the two figures below has the lesser volume?

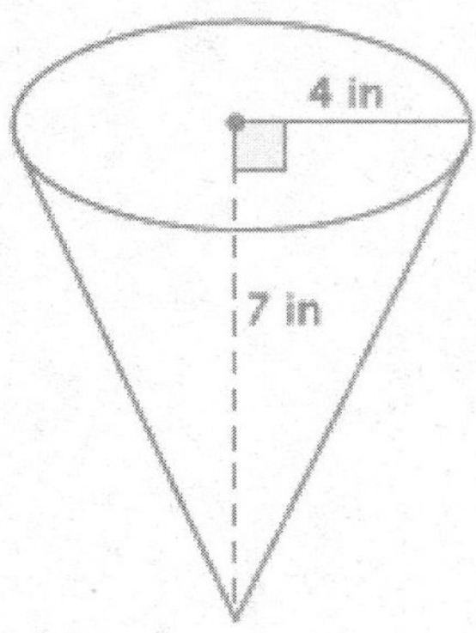


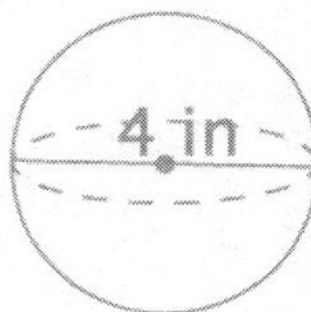


5. Which of the two figures below has the greater volume?

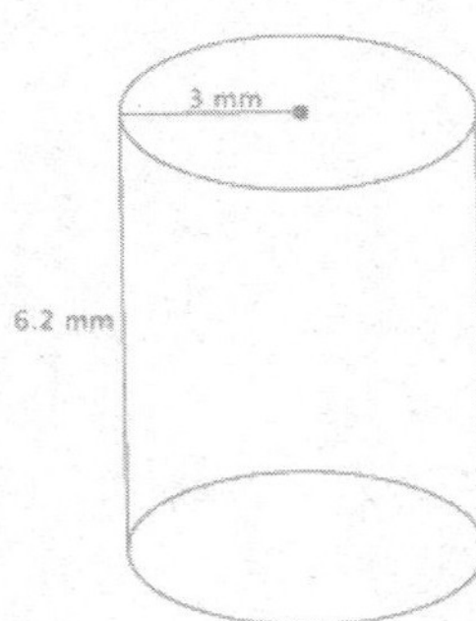


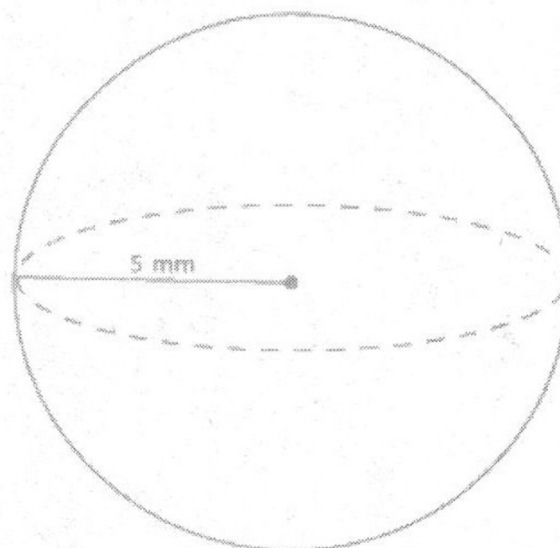


6. Bridget wants to determine which ice cream option is the best choice. The chart below gives the description and prices for her options. Use the space below each item to record your findings.

| \$2.00 | \$3.00 | \$4.00 |
|---|---|---|
| One scoop in a cup | Two scoops in a cup | Three scoops in a cup |
| | | |
| Half a scoop on a cone filled with ice cream | | A cup filled with ice cream (level to the top of the cup) |
| | | |

A scoop of ice cream is considered a perfect sphere and has a 2-inch diameter. A cone has a 2-inch diameter and a height of 4.5 inches. A cup, considered a right circular cylinder, has a 3-inch diameter and a height of 2 inches.

a. Determine the volume of each choice. Use 3.14 to approximate $\pi$.

b. Determine which choice is the best value for her money. Explain your reasoning.

# Credits

Great Minds® has made every effort to obtain permission for the reprinting of all copyrighted material. If any owner of copyrighted material is not acknowledged herein, please contact Great Minds for proper acknowledgment in all future editions and reprints of this module.